Hippocampus - More than Just Memory

Edited by Douglas D. Burman

Published in London, United Kingdom

Hippocampus - More than Just Memory
http://dx.doi.org/10.5772/intechopen.106303
Edited by Douglas D. Burman

Contributors
Lawrence Dayo Adedayo, Gideon Ojo, Samuel Umanah, Ileri-Oluwa Emmanuel, Gideon Aitokhuehi, Olubayode Bamidele, Cecilia Beatriz Moreno, Ángel Javier Tabullo, Donald G. MacKay, Tomokatsu Hori, Hideki Shiramizu, Hajime Miyata, Masoumeh Asadbegi (Hamedi), Alireza Komaki, Azadeh Ebrahim-Habibi, Parsa Amiri, Parichehreh Yaghmaei, Seyed Asaad Karimi, Iraj Salehi, Douglas D. Burman, Livia Livinţ Popa, Diana Chira, Aida Asaftei, Dafin Mureşanu, Stefan Strilciuc, Christiane Neuhaus, André Pedro, Jânio Monteiro, António João Silva

First published in London, United Kingdom, 2023 by IntechOpen
IntechOpen is the global imprint of INTECHOPEN LIMITED, registered in England and Wales, registration number: 11086078, 5 Princes Gate Court, London, SW7 2QJ, United Kingdom

British Library Cataloguing-in-Publication Data
A catalogue record for this book is available from the British Library

Additional hard and PDF copies can be obtained from orders@intechopen.com

Hippocampus - More than Just Memory
Edited by Douglas D. Burman
p. cm.
Print ISBN 978-1-83768-712-1
Online ISBN 978-1-83768-713-8
eBook (PDF) ISBN 978-1-83768-714-5

For EU product safety concerns:
IN TECH d.o.o., Prolaz Marije Krucifikse Kozulić 3, 51000 Rijeka, Croatia,
info@intechopen.com or visit our website at intechopen.com.

Meet the editor

Douglas D. Burman obtained a Ph.D. in Anatomy (Neuroscience) from the University of Illinois Chicago, USA, in 1987. He completed his postgraduate studies at Yale University, Connecticut, USA, and Northwestern University, Illinois, USA. He is currently Director Emeritus of functional MRI (fMRI) services at NorthShore University Health System, Illinois, USA. Dr. Burman is skilled in electrophysiology, cognitive behavior, fMRI (activation, connectivity), and DTI fiber tracking. He has more than fifty scientific publications to his credit.

Contents

Preface IX

Section 1
Introduction 1

Chapter 1 3
Introductory Chapter: A Brief Survey of the Functional Roles of the Hippocampus
by Douglas D. Burman

Section 2
Functional Properties of the Hippocampus 17

Chapter 2 19
Hippocampus: Its Role in Relational Memory
by Lawrence Adedayo, Gideon Ojo, Samuel Umanah, Gideon Aitokhuehi, Ileri-Oluwa Emmanuel and Olubayode Bamidele

Chapter 3 35
Perspective Chapter: Role of the Hippocampal Formation in Navigation from a Simultaneous Location and Mapping Perspective
by André Pedro, Jânio Monteiro and António João Silva

Section 3
Hippocampal Role in Skill Acquisition 59

Chapter 4 61
The Fundamental Role of Memory Systems in Children's Writing Skills
by Cecilia Beatriz Moreno and Ángel Javier Tabullo

Chapter 5 79
Performing Music on Stage: The Role of the Hippocampus in Expert Memory and Culture
by Christiane Neuhaus

Section 4
Hippocampal Plasticity **105**

Chapter 6 **107**
The Modulation of Hippocampus Plasticity
by Livia Livinț Popa, Diana Chira, Aida Asaftei, Stefan Strilciuc and Dafin Mureșanu

Chapter 7 **135**
Learning and Memory Impairment Induced by Amyloid Beta Peptide and Effects of Thymol on Hippocampal Synaptic Plasticity in Rats Fed a High-Fat Diet That Received Amyloid Beta
by Masoumeh Asadbegi (Hamedi), Alireza Komaki, Parsa Amiri, Seyed Asaad Karimi, Parichehreh Yaghmaei, Azadeh Ebrahim-Habibi and Iraj Salehi

Section 5
Effects of Hippocampal Lesions **153**

Chapter 8 **155**
Subtemporal Multiple Hippocampal Transection with/without CA1-Subiculum Disconnection for Medically Intractable Temporal Lobe Epilepsy
by Tomokatsu Hori, Hideki Shiramizu and Hajime Miyata

Chapter 9 **181**
Beyond Memory in H.M., The World's Most Famous "Hippocampal Amnesic"
by Donald G. MacKay

Preface

The hippocampus is undoubtedly the most-studied brain structure in history. Its cytoarchitecture, anatomical connections, neurochemistry and pharmacological responses, in vivo and in vitro electrophysiology across varying states of alertness and task conditions, and behavioral deficits following lesions have been studied throughout the vertebrate kingdom. With the development of neuroimaging techniques, hippocampal response properties and influences on other brain regions during a wide variety of cognitive tasks have been explored in even greater depth in humans.

Following a famous surgical case in the middle of the twentieth century (patient "HM" of Scoville and Milner), the hippocampus has been widely recognized as essential for the formation of declarative and episodic memories, characterized by the subject's awareness of a unique experience. The role of the hippocampus in creating a spatial map of the environment was postulated two decades later based on the pattern of activity of its cells as rats navigated a maze; additional studies confirmed the role of the hippocampus in navigation in other species, including humans. A hippocampal role in both memory and navigation is now widely accepted. Other studies have suggested additional cognitive functions for the hippocampus; because cognitive activities generally require access to memories, however, any additional role for the hippocampus in cognition is more controversial. Factors such as aging, chronic stress, and post-traumatic stress disorder (PTSD) elicit changes in neurochemistry, volume, and/or neural function in the hippocampus, accompanied by observable decreases in cognitive function.

This book explores the functionality of the hippocampus in memory and other cognitive functions. In Section 1, Chapter 1, "Introductory Chapter: A Brief Survey of the Functional Roles of the Hippocampus", briefly reviews cognitive properties reported for the hippocampus; whereas most neural properties may be ascribed to memory, the author concludes that its functionality extends far beyond.

Section 2 further explores the functional properties of the hippocampus. Chapter 2, "Hippocampus: Its Role in Relational Memory", shows how the hippocampus can create connections between different components of an event through its extensive interconnections with diverse cortical regions. Chapter 3, "Perspective Chapter: Role of the Hippocampal Formation in Navigation from a Simultaneous Location and Mapping Perspective", draws parallels between neuronal interactions within the hippocampal formation with the methods used for navigating from satellite data in the Global Positioning System (GPS).

Section 3 characterizes the roles of the hippocampus during skill acquisition. Chapter 4, "The Fundamental Role of Memory Systems in Children's Writing Skills", describes writing ability in children, which develops as the hippocampus and other systems integrate knowledge from long-term memory with cognitive control over various cognitive, phonological, and motor systems. Chapter 5, "Performing Music on Stage: The Role of the Hippocampus in Expert Memory and Culture", explores the hippocampal system and its sensitivity to temporal details for the acquisition of expert performance of music, which plays a role in cultural transmission across centuries as well as entertainment.

Section 4 explores the functionality of hippocampal plasticity. Chapter 6, "The Modulation of Hippocampus Plasticity", discusses how neurogenesis in the hippocampal formation is affected by natural brain chemicals, pharmacological drugs, and non-pharmacological interventions, affecting hippocampal connectivity with areas involved in motor and cognitive control. Chapter 7, "Learning and Memory Impairment Induced by Amyloid Beta Peptide and Effects of Thymol on Hippocampal Synaptic Plasticity in Rats Fed a High-Fat Diet That Received Amyloid Beta" examines how amyloid-beta peptide decreases hippocampal synaptic plasticity, impairing learning and memory, and how these effects can be reduced in rats that are fed thymol.

In Section 5, lesion studies provide clues about essential functions and the functional organization of the hippocampus. Chapter 8, "Subtemporal Multiple Hippocampal Transection with/without CA1-Subiculum Disconnection for Medically Intractable Temporal Lobe Epilepsy", describes how both the presence and magnitude of memory deficits following unilateral hippocampal lesions in epilepsy patients depend on the surgical approach and hippocampal regions excised. IQ deficits in these patients do not perfectly mirror memory deficits, suggesting the hippocampus may be involved in additional cognitive processes. Finally, Chapter 9, "Beyond Memory in H.M., The World's Most Famous "Hippocampal Amnesic"", presents a detailed analysis of experimental data from patient "HM," demonstrating a pattern of cognitive deficits in tasks that do not require memory. A pattern of deficits in understanding non-mnemonic cognitive relationships were observed across a variety of tasks, including those invoking semantics, visual scenes, humor, and integrating familiar concepts into useful ideas.

The chapters in this book clearly demonstrate that the functionality of the hippocampus extends beyond the formation and recall of episodic memories.

Douglas D. Burman
Director Emeritus of fMRI Services,
Department of Radiology,
NorthShore University HealthSystem,
Evanston, Illinois, USA

Section 1

Introduction

Chapter 1

Introductory Chapter: A Brief Survey of the Functional Roles of the Hippocampus

Douglas D. Burman

1. Introduction

Two views of hippocampal function have dominated the literature for the past half-century, each of which has experimental support. One view asserts that the hippocampus is primarily involved in some forms of memory, particularly episodic and long-term declarative memories (i.e., those that can be verbalized). This view was originally based upon the outcome of a surgery in the 1950s, when Scoville and Milner surgically resected a patient's hippocampus to relieve epileptic seizures [1]; the hippocampus is critical for these forms of memory, as confirmed in subsequent human and animal studies.

A second view of hippocampal function is that it is involved in navigation by cognitively mapping our surroundings. This view followed the discovery of hippocampal place cells in rats [2]. The intensity of these cells' activity depended on the animal's location within a baited maze; their relevance to behavior was demonstrated through deficits in navigation following hippocampal lesions [3, 4]. The human hippocampus is similarly involved in navigation [5–7].

These two views of hippocampal function are not mutually exclusive; memory of the spatial location of stimuli is essential for navigation, and spatial locations provide contextual information that is embedded into memories. Whether these properties of the hippocampus represent different functions or different facets of the same function has been debated [8–11].

With the advent of *in vivo* neuroimaging techniques, the hippocampus has been implicated in additional cognitive functions, with hippocampal dysfunction observed in a variety of neurological disorders. This chapter provides brief survey of these findings.

2. Learning and memory

A hippocampal role in learning and memory was identified in the mid-twentieth century from lesion studies. Later in the century, neuroimaging methods were developed to examine differential activity between tasks that differed in memory or other cognitive function, including methods to identify neural activity during individual trials [12]. Regional increases in hippocampal activity were observed during learning and recall, and greater activity was observed during learning trials when the presented stimulus was later recalled successfully [13, 14]. The development of connectivity methods additionally allowed the influence of the hippocampus on other brain regions to be characterized during different types of memory, both during memory acquisition (learning) and recall.

2.1 Learning (memory acquisition)

Different types of memory have been identified based on the conditions in which they occur and the duration of time they may be recalled. Activity in the hippocampus has been observed for many of them; however, the hippocampus is recognized as essential or intimately involved in the acquisition of three of these: episodic, spatial, and motor memory.

2.1.1 Episodic memory

Episodic memory involves the formation and recall of a one-time unique event in one's experiences. The hippocampus is essential for the formation of new episodic memories [15]; however, the hippocampus and prefrontal cortex are jointly involved in the formation as well as recall of episodic memory, with bidirectional flow of information between them [16, 17].

In episodic memory formation, the anterior parahippocampal gyrus (entorhinal and perirhinal cortices) encodes a single item, whereas the hippocampus encodes the relation between stimuli; relevant contextual information is selectively activated in the parahippocampal cortex [18]. A similar pattern of activation is found during episodic memory retrieval.

2.1.2 Spatial memory in navigation

Consistent with its proposed function as a cognitive spatial map [2, 19], hippocampal activity during virtual navigation experiments has been correlated with spatial cues [20–22]. The hippocampal map of space is primarily allocentric [6, 23–27], although other frameworks have also been reported [28, 29]. The hippocampal allocentric map interacts with brain regions that use an egocentric framework. Navigational strategies specify from memory the locations of sensory reference points, suggesting the hippocampal system organizes relational experiences in memory [30]. Functional interactions between the hippocampus and prefrontal cortex occur during goal-directed navigation [31].

2.1.3 Motor learning and memory

The hippocampus is involved in motor learning [32–34] and motor memory consolidation [35]. Connectivity analysis shows the hippocampus primarily influencing the contralateral somatosensory cortex during motor learning, differing from the pattern observed during paced repetitive movements in the absence of motor learning [36].

2.2 Memory recall

Hippocampal properties are consistent with its role in memory recall. Pattern separation and pattern completion are two such properties, localized in different regions of the hippocampus.

Pattern separation transforms similar sensory representations or memories into highly dissimilar representations, distinct from each other; this transformation occurs in the hippocampus, whose pattern of activity differs from that of its inputs [37]. Activity associated with pattern separation is most pronounced in the posterior hippocampus [38], particularly in the CA3 and dentate gyrus regions [39–42]. By contrast, pattern completion reflects expectations about what distinct stimuli are likely to appear based on prior experience. The anterior hippocampus is preferentially involved in pattern completion [38]. The CA3 region supports processes involved in spatial pattern completion (as well as pattern separation), spatial pattern association, novelty detection, and short-term memory [40]. Pattern completion in the hippocampus has been linked to predictive coding in the visual cortex [43].

Some hippocampal neurons fire at successive moments during temporally structured experiences, thus representing the flow of time during specific memories; these have been dubbed 'time cells'. Time cell properties parallel those of hippocampal place cells, providing an additional dimension to be integrated with spatial mapping [44]. The CA1 region supports processes associated with temporal pattern completion, temporal pattern association and intermediate-term memory [40].

Recollection as a form of memory recall is distinct from familiarity. Studies of humans, monkeys, and rats using multiple techniques suggest that the hippocampus is critical for recollection but not familiarity [45]. Recollection can be triggered by a cue that shares one or more elements with the original memory, so by using a cue, the chronology of memory recall can be studied. Pattern completion in the hippocampus begins 500 ms after cue onset, triggering the reinstatement of the target memory in neocortex between 500 and 1500 ms; this gives rise to the subjective feeling of recollection [46]. This process engages temporal dynamics, including the reversal of perceptual processing streams and clocking by theta rhythms.

3. Context and cognition

Hippocampal properties include sensitivity to temporal and spatial relationships [47–49], which play a role in scene perception and reconstruction [50]. Some interpret these properties as contextual elements required for memory recall [9, 44, 47]; others suggest a more fundamental perceptual role, which may consequently be incorporated into memories [51, 52]. As evident from the examples below, differences between these viewpoints are often nuanced.

According to one viewpoint [53], space and time break up experiences into specific contexts; these features help organize multimodal inputs. If relevant, additional dimensions (such as emotions) can also be incorporated into an event-defined context. Conceiving of hippocampal representations as constrained by task demands, this viewpoint attempts to unify disparate findings on hippocampal representations of space, time, and other dimensions on its core function.

Another theory describes a prefrontal-hippocampal comparator for voluntary action [54, 55]. Action plans are elaborated by the prefrontal cortex, and serve to guide goal-directed behavior. The prefrontal cortex initiates its plan by transmitting an "efference copy" (corollary discharge) to the CA1 region of the hippocampus, which stores it in working memory. This efference copy includes the expected outcomes of the action plan, including the personal and subjective experience of the intended behavior, when, and in what context. The CA1 region of the hippocampus compares the response intention with the actual outcome through cortical interactions mediated through the hippocampal theta rhythm; the theta power serves as a prediction error signal during hippocampal dependent tasks. When a mismatch occurs, an error signal in the hippocampus is transmitted to the prefrontal cortex, the action plan is reformulated, and working memory is updated. When the expected and actual outcomes match, the hippocampus transmits a signal to strengthen or consolidate the action plan in prefrontal cortex.

The hippocampus and the ventromedial prefrontal cortex interact when making decisions, integrating episodic memory via the hippocampus with value-based decision-making via the ventromedial prefrontal cortex [56]. The anterior parts of the hippocampus in humans (the ventral hippocampus in rodents) may also contribute to approach-avoidance conflict decision-making [57]. Such a scenario arises when a goal stimulus is simultaneously associated with reward and punishment.

The hippocampus has been suggested to play a critical role in behavioral flexibility. Cognition and social behaviors often require flexible use of information that can result from the formation, recombination, and reconstruction of relational memory representations. By filling this function, the hippocampus may play an instrumental role in abilities as diverse as decision-making, character judgments, establishing and maintaining social bonds, empathy, social discourse, navigation, exploration, creativity, imagination, memory, and language use [58].

3.1 Cognitive deficits associated with hippocampal dysfunction

During the progression of Alzheimer's, changes are observed in hippocampal size [59, 60], function [61], and connectivity [62, 63]. Degenerative processes result in the accretion of plaques and tangles; their presence indicates the local loss of neuronal function. Plaques and tangles first appear in the entorhinal cortex, followed shortly thereafter in the hippocampus itself [64–66].

In schizophrenia, connections between the hippocampus and prefrontal regions are dysfunctional, consistent with a dysconnection syndrome [67].

Parkinson's, at least in some cases, may also involve the hippocampus. A gene that contributes to familial and juvenile Parkinsonism disrupts hippocampal synaptic transmission in vitro [68]; if the hippocampus plays a role in cognitive control [69], this could result in motor dysfunction. The hippocampus has also been implicated in the cognitive dysfunction observed in some Parkinson's patients [70].

Cognitive effects of concussions may also arise, at least in part, from damage to the hippocampal system [71, 72]. During concussions, blunt force to the skull typically generates torsional forces that damage the brain. The greatest torsional forces appear in the interior of the skull, including the corpus callosum and nearby structures such as the hippocampus [73–75]. These forces stretch nerve fibers in the region, resulting in necrosis.

Alterations in hippocampus size or function have also been noted in other neurological conditions, including depression [76, 77], ADHD [78, 79], and PTSD [80, 81].

3.2 Neurogenesis and environmental enrichment

With the proliferation and differentiation of adult neural stem cells, new neurons are generated throughout adulthood in the subgranular zone of the dentate gyrus in the hippocampus [82]. Adult hippocampal neurogenesis is thought to play a major role in hippocampus-dependent functions. By integrating new neurons with the structural plasticity of mature neurons, adult neurogenesis may maintain hippocampal plasticity in its circuits [83]. Adult-born neurons may play distinct physiological roles in hippocampus-dependent functions, such as memory encoding and mood regulation [84] as well as spatial learning [85]. Exercise has been suggested to increase the production of neurons, whereas environment enrichment increases the likelihood of their survival through cortical restructuring [85]. This restructuring results from a transient increase in cell activity and structural plasticity, which leads to improved cognition [86].

4. Executive function (cognitive control)

As more information about hippocampal activity has accrued, other roles for the hippocampus have also been suggested, including a role in cognitive control [28, 29, 69, 87]. Cognitive control refers to processes that organize different thoughts, separate currently-relevant and irrelevant information, and coordinate thoughts and actions. One report reviewed place cell studies that used experimental manipulations to dissociate the environment into two or more spatial frames of locations, typically to test notions of pattern separation. The ensemble discharge in the hippocampus self-organized into multiple, transiently-organized representations of space; separate representations of frame-specific positions alternated on timescales from 25 ms to several seconds. The dynamic, functional grouping of discharge predicted the animal's behavioral needs, which suggested a hippocampal role in cognitive control [28].

A broader role in cognitive control, perhaps in conjunction with the prefrontal cortex, has been suggested based on a consistent pattern showing increased hippocampal connectivity with whichever cortical areas are required for task performance [69]. During a volitional movement task, for example, the specificity of this connectivity was pronounced, with hippocampal connectivity linked spatiotemporally to the representation of the moving finger [87].

5. Consciousness

The hippocampus has been suggested to play a role in conscious perception by integrating information that identifies an object with its spatiotemporal location, embedded within an emotional context [88–90]. In this view, the hippocampus acts via the medial prefrontal cortex, influencing prefrontal top-down attentional control of sensory processing and thus event memory formation. Citing specific functional deficits in the hippocampus and its neurochemical connections with prefrontal cortex, the authors suggest that weakly-related sensory representations within the hippocampus underlie hallucinations in schizophrenia [88].

Recently, the hippocampus has been proposed to play a central role in all features associated with the normal alert state of consciousness [91]. This state is defined by cognitive characteristics associated with consciousness in a neurologically-intact individual who is awake and alert; this includes sensory perceptions, learning, memory, attention, language, thoughts, emotional responsiveness, decision-making, and motor control. Using evidence from connectivity analyses, Burman demonstrates hippocampal influences on relevant cortical areas involved in all these cognitive processes. Details of his model (and some of his findings) reflect the joint influence from homotopic regions of the left and right hippocampus, the consistent finding that hippocampal connectivity increases in regions that execute a task, and its influence on other regions (especially prefrontal cortex and the precuneus) that support task performance. By recalling memories of self-experiences across periods of sleep and earlier periods of one's life, the hippocampus is also noted to show characteristics consistent with the conscious sense of self.

6. Hippocampus: one function or many?

The hippocampus has been implicated in many cognitive functions and disorders. One question that arises is whether the hippocampus is functionally diverse, perhaps with different regions specialized for different functions, or whether the various functions ascribed to the hippocampus may all result from its known role in memory. Tasks that are not intended to involve memory almost invariably require a mnemonic component; even a repetitive tapping task that avoids motor learning [36] requires a subject to remember what behavior is required. Hippocampal properties that do not reflect the traditional relationship to memory or learning, such as sensitivity to time or spatial relationships, may reflect contextual elements that are essential for memory [9, 44, 49, 51]. If all properties of the hippocampus are consistent with the requirements for creating and recalling memories, however, the *functionality* of the hippocampus extends beyond our traditional view of memory, as the hippocampus is involved in decision-making, navigation, intentional movements, language, and many other cognitive functions.

Improvements in experimental methods have helped elucidate the hippocampal role in such diverse functions. Effective connectivity analysis, for example, examines the directional influence of one brain area on another; using this tool, hippocampal influences are observed to be selective for those brain areas relevant for task behavior. Depending on the task, the hippocampus inversely influences activity in sensorimotor cortex for motor tasks, the ventral occipital color region for tasks requiring attention to color (Stroop task and a conjunction task requiring attention to both color and shape), auditory association cortex when presented with annoying sounds, the temporal phonology region when making a rhyming judgment, and the temporal semantic region when making semantic judgments [69]. These influences do not reflect the activation of a specific memory, and thus is inconsistent with a direct role in recall; rather, they reflect the collective influence of memories through the hippocampus on processing in sensory and motor areas of the brain.

Similarly, the hippocampus shows bidirectional interactions with prefrontal cortex during tasks that require planning and memory recall [5, 92, 93]. The hippocampus coordinates brain functions associated with cognitive functions through theta waves and cross-frequency coupling [94–101]. Hippocampal functionality often appears to be mediated, directly or indirectly, through other brain regions.

Involvement of the hippocampus in diverse cognitive processes thus results from extensive interactions with other brain regions depending on the task at hand. Although the *functional properties* of the hippocampus may be described in relationship to memory, the *functionality* of the hippocampus goes far beyond.

7. Conclusion

The hippocampus is best known for its essential role in episodic, declarative long-term memories, yet a survey of findings indicates a much broader role in cognition. Hippocampal influences on other brain regions are extensive, but appear to be specific to the cognitive requirements of the task at hand.

Author details

Douglas D. Burman
Northshore University HealthSystem, Evanston, Illinois, USA

*Address all correspondence to: ddburman73@yahoo.com

References

[1] Scoville WB, Milner B. Loss of recent memory after bilateral hippocampal lesions. Journal of Neurology, Neurosurgery, and Psychiatry. 1957;**20**(1):11-21

[2] O'Keefe J, Dostrovsky J. The hippocampus as a spatial map: Preliminary evidence from unit activity in the freely-moving rat. Brain Research. 1971;**34**(1):171-175

[3] Hollup SA, Kjelstrup KG, Hoff J, Moser M-B, Moser EI. Impaired recognition of the goal location during spatial navigation in rats with hippocampal lesions. The Journal of Neuroscience. 2001;**21**(12):4505-4513

[4] Whishaw IQ, McKenna JE, Maaswinkel H. Hippocampal lesions and path integration. Current Opinion in Neurobiology. 1997;**2**(7):228-234

[5] Brown TI, Carr VA, LaRocque KF, Favila SE, Gordon AM, Bowles B, et al. Prospective representation of navigational goals in the human hippocampus. Science. 2016;**352**(6291):1323-1326

[6] Igloi K, Doeller CF, Berthoz A, Rondi-Reig L, Burgess N. Lateralized human hippocampal activity predicts navigation based on sequence or place memory. Proceedings of the National Academy of Sciences. 2010;**107**(32):14466-14471

[7] Maguire EA, Burgess N, Donnett JG, Frackowiak RS, Frith CD, O'Keefe J. Knowing where and getting there: A human navigation network. [see comment]. Science. 1998;**280**(5365):921-924

[8] Eichenbaum H, Cohen NJ. Can we reconcile the declarative memory and spatial navigation views on hippocampal function? Neuron. 2014;**83**(4):764-770

[9] Eichenbaum H. On the integration of space, time, and memory. Neuron. 2017;**95**(5):1007

[10] Rolls ET. Functions of the Primate Hippocampus in Spatial Processing and Memory: Neurobiology of Comparative Cognition. New York: Psychology Press; 2014. pp. 359-382

[11] Smith DM, Mizumori SJY. Hippocampal place cells, context, and episodic memory. Hippocampus. 2006;**16**(9):716-729

[12] Hopfinger JB, Büchel C, Holmes AP, Friston KJ. A study of analysis parameters that influence the sensitivity of event-related fMRI analyses. NeuroImage. 2000;**11**(4):326-333

[13] Hannula DE, Ranganath C. Medial temporal lobe activity predicts successful relational memory binding. The Journal of Neuroscience. 2008;**28**(1):116-124

[14] Jenkins LJ, Ranganath C. Prefrontal and medial temporal lobe activity at encoding predicts temporal context memory. The Journal of Neuroscience. 2010;**30**(46):15558-15565

[15] Spiers HJ, Maguire EA, Burgess N. Hippocampal amnesia. Neurocase. 2001;**7**(5):357-382

[16] Rugg MD, Otten LJ, Henson RN. The neural basis of episodic memory: Evidence from functional neuroimaging. Philosophical Transactions of the Royal Society of London. Series B, Biological Sciences. 2002;**357**(1424):1097-1110

[17] Eichenbaum H. Prefrontal-hippocampal interactions in episodic

memory. Nature Reviews. Neuroscience. 2017;**18**(9):547-558

[18] Tsukiura T. Functional neuroimaging studies of episodic memory--functional dissociation in the medial temporal lobe structures. Brain & Nerve / Shinkei Kenkyu no Shinpo. 2008;**60**(7):833-844

[19] O'Keefe J, Nadel L. The Hippocampus as a Cognitive Map. Oxford: Oxford University Press; 1978

[20] Hirshhorn M, Grady C, Rosenbaum RS, Winocur G, Moscovitch M. The hippocampus is involved in mental navigation for a recently learned, but not a highly familiar environment: A longitudinal fMRI study. Hippocampus. 2012;**22**(4):842-852

[21] Ohnishi T, Matsuda H, Hirakata M, Ugawa Y. Navigation ability dependent neural activation in the human brain: An fMRI study. Neuroscience Research. 2006;**55**(4):361-369

[22] Rodriguez PF. Neural decoding of goal locations in spatial navigation in humans with fMRI. Human Brain Mapping. 2010;**31**(3):391-397

[23] Albouy G, Fogel S, King BR, Laventure S, Benali H, Karni A, et al. Maintaining vs. enhancing motor sequence memories: Respective roles of striatal and hippocampal systems. NeuroImage. 2015;**108**:423-434

[24] Holdstock J, Mayes A, Cezayirli E, Isaac C, Aggleton JP, Roberts N. A comparison of egocentric and allocentric spatial memory in a patient with selective hippocampal damage. Neuropsychologia. 2000;**38**(4):410-425

[25] Rolls ET. Hippocampal spatial view cells for memory and navigation, and their underlying connectivity in humans. Hippocampus. 2022:1-40. DOI: 10.1002/hipo.23467

[26] Rolls ET, Wirth S. Spatial representations in the primate hippocampus, and their functions in memory and navigation. Progress in Neurobiology. 2018;**171**:90-113

[27] Zaehle T, Jordan K, Wüstenberg T, Baudewig J, Dechent P, Mast FW. The neural basis of the egocentric and allocentric spatial frame of reference. Brain Research. 2007;**1137**:92-103

[28] Kelemen E, Fenton AA. Coordinating different representations in the hippocampus. Neurobiology of Learning and Memory. 2016;**129**:50-59

[29] Kelemen E, Fenton AA. Dynamic grouping of hippocampal neural activity during cognitive control of two spatial frames. PLoS Biology. 2010;**8**(6):e1000403

[30] Eichenbaum H. The role of the hippocampus in navigation is memory. Journal of Neurophysiology. 2017;**117**(4):1785

[31] Ito HT. Prefrontal–hippocampal interactions for spatial navigation. Neuroscience Research. 2018;**129**:2-7

[32] Genzel L, Dresler M, Cornu M, Jäger E, Konrad B, Adamczyk M, et al. Medial prefrontal-hippocampal connectivity and motor memory consolidation in depression and schizophrenia. Biological Psychiatry. 2015;**77**(2):177-186

[33] Fernandez-Seara MA, Aznarez-Sanado M, Mengual E, Loayza FR, Pastor MA. Continuous performance of a novel motor sequence leads to highly correlated striatal and

hippocampal perfusion increases. NeuroImage. 2009;**47**(4):1797-1808

[34] Ramnani N, Toni I, Josephs O, Ashburner J, Passingham RE. Learning- and expectation-related changes in the human brain during motor learning. Journal of Neurophysiology. 2000;**84**(6):3026-3035

[35] Walker MP, Stickgold R, Alsop D, Gaab N, Schlaug G. Sleep-dependent motor memory plasticity in the human brain. Neuroscience. 2005;**133**(4):911-917

[36] Burman DD. Hippocampal connectivity with sensorimotor cortex during volitional finger movements: Laterality and relationship to motor learning. PLoS One. 2019;**14**(9):e0222064

[37] Kirwan CB, Stark CEL. Overcoming interference: An fMRI investigation of pattern separation in the medial temporal lobe. Learning & Memory. 2007;**14**(9):625-633

[38] Poppenk J, Evensmoen HR, Moscovitch M, Nadel L. Long-axis specialization of the human hippocampus. Trends in Cognitive Sciences. 2013;**17**(5):230-240

[39] Bakker A, Kirwan CB, Miller M, Stark CEL. Pattern separation in the human hippocampal CA3 and dentate gyrus. Science. 2008;**319**(5870):1640-1642

[40] Kesner RP, Lee I, Gilbert P. A behavioral assessment of hippocampal function based on a subregional analysis. Reviews in the Neurosciences. 2004;**15**(5):333

[41] Yassa MA, Lacy JW, Stark SM, Albert MS, Gallagher M, Stark CEL. Pattern separation deficits associated with increased hippocampal CA3 and dentate gyrus activity in nondemented older adults. Hippocampus. 2011;**21**(9):968-979

[42] Rolls ET. Pattern separation, completion, and categorisation in the hippocampus and neocortex. Neurobiology of Learning and Memory. 2016;**129**:4-28

[43] Hindy NC, Ng FY, Turk-Browne NB. Linking pattern completion in the hippocampus to predictive coding in visual cortex. Nature Neuroscience. 2016;**19**(5):665-667

[44] Eichenbaum H. Time cells in the hippocampus: A new dimension for mapping memories. Nature Reviews. Neuroscience. 2014;**15**(11):732-744

[45] Eichenbaum H, Yonelinas AP, Ranganath C. The medial temporal lobe and recognition memory. Annual Review of Neuroscience. 2007;**30**:123-152

[46] Staresina BP, Wimber M. A neural chronometry of memory recall. Trends in Cognitive Sciences. 2019;**23**(12):1071-1085

[47] Barnett AJ, O'neil EB, Watson HC, Lee ACH. The human hippocampus is sensitive to the durations of events and intervals within a sequence. Neuropsychologia. 2014;**64**:1-12

[48] Libby LA, Hannula DE, Ranganath C. Medial temporal lobe coding of item and spatial information during relational binding in working memory. The Journal of Neuroscience. 2014;**34**:14233-14242

[49] Olsen RK, Moses SN, Riggs L, Ryan JD. The hippocampus supports multiple cognitive processes through relational binding and comparison. Frontiers in Human Neuroscience. 2012;**6**:146

[50] Zeidman P. Mullally SaL, Maguire EA. Constructing, perceiving, and maintaining scenes: Hippocampal activity and connectivity. Cerebral Cortex. 2015;**25**(10):3836-3855

[51] Howard MW, Eichenbaum H. Time and space in the hippocampus. Brain Research. 1621;**2015**:345-354

[52] MacDonald CJ. Prospective and retrospective duration memory in the hippocampus: Is time in the foreground or background? Philosophical Transaction Royal Society B. 2014;**369**(1637):20120463

[53] Ekstrom AD, Ranganath C. Space, time, and episodic memory: The hippocampus is all over the cognitive map. Hippocampus. 2017;**28**(9):680-687. DOI: 10.1002/hipo.22750

[54] Numan R. A prefrontal-hippocampal comparator for goal-directed behavior: The intentional self and episodic memory. Frontiers in Behavioral Neuroscience. 2015;**9**:323

[55] Numan R. The Prefrontal-Hippocampal Comparator: Volition and Episodic Memory. Perceptual and Motor Skills. 2021;**128**(6):2421-2447

[56] Weilbächer RA, Gluth S. The interplay of hippocampus and ventromedial prefrontal cortex in memory-based decision making. Brain Sciences. 2017;**7**(1):E4. DOI: 10.3390/brainsci7010004

[57] Ito R, Lee ACH. The role of the hippocampus in approach-avoidance conflict decision-making: Evidence from rodent and human studies. Behavioural Brain Research. 2016;**313**:345-357

[58] Rubin RD, Watson PD, Duff MC, Cohen NJ. The Role of the Hippocampus in Flexible Cognition and Social Behavior. Frontiers in Human Neuroscience. 2014;**8**:742

[59] Morra JH, Tu Z, Apostolova LG, Green AE, Avedissian C, Madsen SK, et al. Automated mapping of hippocampal atrophy in 1-year repeat MRI data from 490 subjects with Alzheimer's disease, mild cognitive impairment, and elderly controls. NeuroImage. 2009;**45**(1):S3-S15

[60] Pennanen C, Kivipelto M, Tuomainen S, Hartikainen P, Hanninen T, Laakso MP, et al. Hippocampus and entorhinal cortex in mild cognitive impairment and early AD. Neurobiology of Aging. 2004;**25**(3):303-310

[61] Sperling R. Functional MRI studies of associative encoding in normal aging, mild cognitive impairment, and Alzheimer's disease. Annals of the New York Academy of Sciences. 2007;**1097**:146-155

[62] Wang L, Zang Y, He Y, Liang M, Zhang X, Tian L, et al. Changes in hippocampal connectivity in the early stages of Alzheimer's disease: Evidence from resting state fMRI. NeuroImage. 2006;**31**(2):496-504

[63] Zhang H-Y, Wang S-J, Liu B, Ma Z-L, Yang M, Zhang Z-J, et al. Resting Brain Connectivity: Changes during the Progress of Alzheimer Disease 1. Radiology. 2010;**256**(2):598-606

[64] Van Hoesen GW, Hyman BT, Damasio AR. Entorhinal cortex pathology in Alzheimer's disease. Hippocampus. 1991;**1**(1):1-8

[65] Coutureau E, Di Scala G. Entorhinal cortex and cognition. Progress in Neuro-Psychopharmacology & Biological Psychiatry. 2009;**33**(5):753-761

[66] Gómez-Isla T, Price JL, McKeel DW Jr, Morris JC, Growdon JH, Hyman BT.

Profound loss of layer II entorhinal cortex neurons occurs in very mild Alzheimer's disease. The Journal of Neuroscience. 1996;**16**(14):4491-4500

[67] Woodcock EA, Wadehra S, Diwadkar VA. Network profiles of the dorsal anterior cingulate and dorsal prefrontal cortex in schizophrenia during hippocampal-based associative memory. Frontiers in Systems Neuroscience. 2016;**10**:32

[68] Zhu M, Cortese GP, Waites CL. Parkinson's disease-linked Parkin mutations impair glutamatergic signaling in hippocampal neurons. BMC Biology. 2018;**16**(1):1-20

[69] Burman DD. Hippocampal Influences on Movements, Sensory, and Language Processing: A Role in Cognitive Control?: Hippocampus-Cytoarchitecture and Diseases. London, UK: IntechOpen; 2021. pp. 57-79

[70] Calabresi P, Castrioto A, Di Filippo M, Picconi B. New experimental and clinical links between the hippocampus and the dopaminergic system in Parkinson's disease. The Lancet Neurology. 2013;**12**(8):811-821

[71] Brett BL, Savitz J, Nitta M, España L, Teague TK, Nelson LD, et al. Systemic inflammation moderates the association of prior concussion with hippocampal volume and episodic memory in high school and collegiate athletes. Brain, Behavior, and Immunity. 2020;**89**:380-388

[72] Meier TB, España LY, Kirk AJ, Nader AM, Powell JE, Nelson LD, et al. Association of previous concussion with hippocampal volume and symptoms in collegiate-aged athletes. Journal of Neurotrauma. 2021;**38**(10):1358-1367

[73] Martin G. Traumatic brain injury: The first 15 milliseconds. Brain Injury. 2016;**30**(13-14):1517-1524

[74] Bigler ED. Neuropsychology and clinical neuroscience of persistent post-concussive syndrome. Journal of the International Neuropsychological Society. 2008;**14**(1):1-22

[75] Ng LJ, Volman V, Gibbons MM, Phohomsiri P, Cui J, Swenson DJ, et al. A mechanistic end-to-end concussion model that translates head kinematics to neurologic injury. Frontiers in Neurology. 2017;**8**:269

[76] Fotuhi M, Do D, Jack C. Modifiable factors that alter the size of the hippocampus with ageing. Nature Reviews. Neurology. 2012;**8**(4):189-202

[77] Frodl T, Schaub A, Banac S, Charypar M, Jager M, Kummler P, et al. Reduced hippocampal volume correlates with executive dysfunctioning in major depression. Journal of Psychiatry & Neuroscience. 2006;**31**(5):316

[78] Rangarajan B, Suresh S, Mahanand B. Identification of potential biomarkers in the hippocampus region for the diagnosis of ADHD using PBL-McRBFN approach. In: 2014 13th International Conference on Control Automation Robotics & Vision (ICARCV). Singapore: IEEE; 2014

[79] Plessen KJ, Bansal R, Zhu H, Whiteman R, Amat J, Quackenbush GA, et al. Hippocampus and amygdala morphology in attention-deficit/ hyperactivity disorder. Archives of General Psychiatry. 2006;**63**(7): 795-807

[80] Liberzon I, Sripada CS. The functional neuroanatomy of PTSD: A critical review. Progress in Brain Research. 2008;**167**:151-169

[81] Malivoire BL, Girard TA, Patel R, Monson CM. Functional connectivity of hippocampal subregions in PTSD: Relations with symptoms. BMC Psychiatry. 2018;**18**(1):129

[82] Boldrini M, Fulmore CA, Tartt AN, Simeon LR, Pavlova I, Poposka V, et al. Human hippocampal neurogenesis persists throughout aging. Cell Stem Cell. 2018;**22**(4):598-599

[83] Christian KM, Song H, Ming G. Functions and dysfunctions of adult hippocampal neurogenesis. Annual Review of Neuroscience. 2014;**37**:243

[84] Toda T, Parylak SL, Linker SB, Gage FH. The role of adult hippocampal neurogenesis in brain health and disease. Molecular Psychiatry. 2019;**24**(1):67-87

[85] Olson AK, Eadie BD, Ernst C, Christie BR. Environmental enrichment and voluntary exercise massively increase neurogenesis in the adult hippocampus via dissociable pathways. Hippocampus. 2006;**16**(3):250-260

[86] Eckert MJ, Abraham WC. Effects of environmental enrichment exposure on synaptic transmission and plasticity in the hippocampus. Neurogenesis and Neural Plasticity. 2013;**2013**:165-187

[87] Burman DD. Hippocampal connectivity with sensorimotor cortex during volitional finger movements. II. Spatial and temporal selectivity. bioRxiv. 2018. 10.1101/479436

[88] Behrendt R-P. Contribution of hippocampal region CA3 to consciousness and schizophrenic hallucinations. Neuroscience and Biobehavioral Reviews. 2010;**34**(8):1121-1136

[89] Behrendt R-P. Hippocampus and consciousness. Reviews in the Neurosciences. 2013;**24**(3):239-266

[90] Behrendt R-P. Conscious experience and episodic memory: Hippocampus at the crossroads. Frontiers in Psychology. 2013;**4**:304

[91] Burman DD. Normal Alert Consciousness: A Central Executive Model of Hippocampal Function. London, UK: InTechOpen [In press]

[92] Dagher A, Owen AM, Boecker H, Brooks DJ. The role of the striatum and hippocampus in planning. Brain. 2001;**124**(5):1020-1032

[93] Kaplan R, King J, Koster R, Penny WD, Burgess N, Friston KJ. The neural representation of prospective choice during spatial planning and decisions. PLoS Biology. 2017;**15**(1):e1002588

[94] Axmacher N, Henseler MM, Jensen O, Weinreich I, Elger CE, Fell J. Cross-frequency coupling supports multi-item working memory in the human hippocampus. Proceedings of the National Academy of Sciences. 2010;**107**(7):3228-3233

[95] Backus AR, Schoffelen J-M, Szebényi S, Hanslmayr S, Doeller CF. Hippocampal-prefrontal theta oscillations support memory integration. Current Biology. 2016;**26**(4):450-457

[96] Battaglia FP, Benchenane K, Sirota A, Pennartz CMA, Wiener SI. The hippocampus: Hub of brain network communication for memory. Trends in Cognitive Sciences. 2011;**15**(7):310-318

[97] Bohbot VD, Copara MS, Gotman J, Ekstrom AD. Low-frequency theta oscillations in the human hippocampus during real-world and virtual navigation. Nature Communications. 2017;**8**:14415

[98] Buzsaki G, Moser EI. Memory, navigation and theta rhythm in the hippocampal-entorhinal system. Nature Neuroscience. 2013;**16**(2):130-138

[99] Herweg NA, Apitz T, Leicht G, Mulert C, Fuentemilla Garriga L,

Bunzeck N. Theta-alpha oscillations bind the hippocampus, prefrontal cortex, and striatum during recollection: Evidence from simultaneous EEG-fMRI. The Journal of Neuroscience. 2016;**36**(12):3579-3587

[100] Piai V, Anderson KL, Lin JJ, Dewar C, Parvizi J, Dronkers NF, et al. Direct brain recordings reveal hippocampal rhythm underpinnings of language processing. Proceedings of the National Academy of Sciences. 2016;**113**(40):11366-11371

[101] Terada S, Sakurai Y, Nakahara H, Fujisawa S. Temporal and rate coding for discrete event sequences in the hippocampus. Neuron. 2017;**94**(6):1248-1262

Section 2

Functional Properties of the Hippocampus

Chapter 2

Hippocampus: Its Role in Relational Memory

Lawrence Adedayo, Gideon Ojo, Samuel Umanah, Gideon Aitokhuehi, Ileri-Oluwa Emmanuel and Olubayode Bamidele

Abstract

Hippocampus is the region of the brain that is primarily associated with memory. The hippocampus, which is located in the inner (medial) region of the temporal lobe, forms part of the limbic system, which is important in regulating emotional responses. The hippocampus is thought to be principally involved in storing long-term memories and in making those memories resistant to forgetting, though this is a matter of debate. It is also thought to play an important role in spatial processing and navigation. Cholinergic system has implicated in the functionality of hippocampus interconnections with other neurons for efficient memory modulation. Pyramidal and globular cells are the main cells of the cornus ammonis and the dentate gyrus which is essential in relational memory consolidation. Acetylcholine is the main neurotransmitter implicated in encoding of memory in the hippocampus. There are diseases that are associated with hippocampus relational memory such as Alzheimer's disease which is currently a global challenge. The hippocampus communicates with widespread regions of cortex through a group of highly interconnected brain regions in the medial temporal lobe. There is paucity of data on its role on relational memory. Therefore, the role of hippocampus in relational memory will be elucidated in this chapter.

Keywords: hippocampus, relational memory, cornus ammonis, acetylcholine, Alzheimer's disease

1. Introduction

The hippocampus was first referred to by a Venetian anatomist Julius Caesar Aranzi in 1587. He described it as a ridge along the floor of the temporal horn of the lateral ventricle and likened first to silkworm, and later to a seahorse (**Figure 1**). In the 1740s, Rene-Jacques Croissant de Garengeot, a Parisian surgeon, coined the term "cornu ammonis," meaning the horn of Amun, and ancient Egyptian god. The organ is coined from two Greek words "hippo" for horse and "kampos" for sea [1].

The hippocampus is the "flash drive" of the human brain and often associated with memory consolidation and decision-making, but it is far more complex in

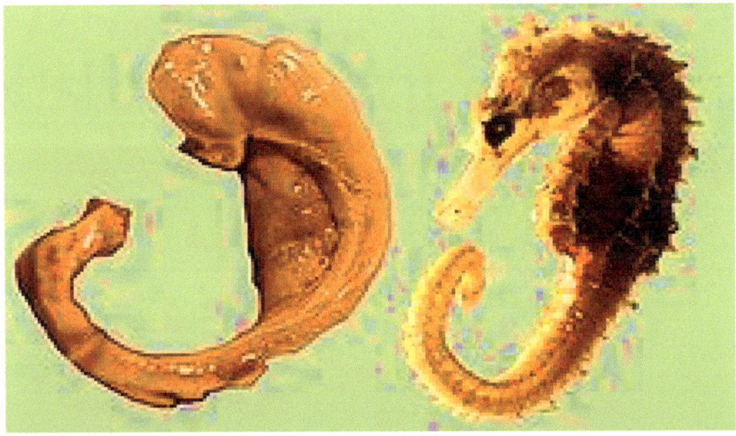

Figure 1.
Seahorse-shaped hippocampus.

structure and function than a flash drive [2]. The hippocampus is a convex elevation of gray matter tissue within the parahippocampal gyrus inside the inferior temporal horn of the lateral ventricle. One can describe it more holistically as a curved and recurved sheet of the cortex that folds into the temporal lobe's medial surface.

2. Anatomical enumeration of hippocampus

2.1 Embryology of hippocampus

In the human embryo, the hippocampal formation develops in relation to the medial surface of each cerebral hemisphere close to the choroid fissure of the lateral ventricle [3]. It is at first approximately, C-shaped in accordance with the outline of the body and inferior horn of the ventricle. The upper part of the formation is, however, separated from the ventricle because of the development of the corpus callosum between the two [4]. For the same reason, this part of the formation remains underdeveloped and is represented by a thin layer of gray matter, lining the upper surface of the corpus callosum. This layer is the indusium griseum. Within the indusium griseum are embedded two bundles of longitudinally running fibers called the medial and lateral longitudinal striae (on each side of the midline). Posteriorly, the indusium griseum is continuous with a thin layer of gray matter related to the inferior aspect of the splenium of the corpus callosum [3].

This gray matter is the splenial gyrus or gyrus fasciolaris. The splenial gyrus runs forward to become continuous with the dentate gyrus, present in relation to the inferior horn of the lateral ventricle. In the region of the inferior horn of the lateral ventricle, the developing hippocampus is pushed into the cavity of the ventricle because of the great development of the neighboring neocortex. The hippocampal formation is best developed in this region and forms the hippocampus. This term includes the dentate gyrus [4].

2.2 Gross anatomy of hippocampus

The hippocampus has three distinct zones: the dentate gyrus, the hippocampus proper, and the subiculum. The dentate gyrus and hippocampus proper form two

C-shaped rings that interlock. The subiculum is thus a transition zone, linking the hippocampus proper with the dentate gyrus [5].

The parahippocampal gyrus and cingulate sulci are located on the medial surface of the hemisphere, forming a C-shaped ring. The medial temporal lobe cortex includes major subdivisions such as the hippocampus and the entorhinal cortex. This five centimeter-long hippocampus (from the anterior end at amygdala to posterior end near the splenium of the corpus callosum) divides into a head, body, and tail [5]. The head is expanded and bears two or three shallow grooves called pes hippocampi. The head of the hippocampus is part of the posterior half of the triangular uncus and is separated inferiorly from the parahippocampal gyrus by the uncal sulcus. The alveus, which is the surface of the hippocampus, is covered by the ependymal tissue inside the ventricular cavity [5].

The fornix, which is the main outflow bundle out of the hippocampus, wraps around the thalamus, where it then becomes separated by the choroidal fissure and the choroid plexus. The hippocampus contains parts like the fimbria, crus, body, and column. The fimbria forms where alveus fibers converge along the medial portion of the lateral ventricle's inferior horn. The white matter of the fimbria separates to form a crux of the ipsilateral fornix at a point beyond the splenium of the corpus callosum.

The anterior choroidal artery runs medially and superiorly to the uncus, between the ambient and semilunar gyrus [5]. It then sends perforating arteries to reach deeper structures. The uncus is closely related to the middle cerebral arteries and its lenticulostriate arteries. The posterior cerebral artery and the basal vein supplies and drains the caudal part of the head of the hippocampus that faces the crus cerebri and crural cistern [6]. Internal cerebral veins drain into thalamostriatal basal ganglia, thalamus, internal capsule, tela choroidea of three ventricles, and hippocampus. The veins on each side unite to form the internal cerebral vein.

2.3 Histology of hippocampus

Stratum Oriens: It is the next layer superficial to the alveus. The cell bodies of inhibitory basket cells and horizontal trilaminar cells, named for their axons innervating three layers, the oriens, Pyramidal, and radiatum, are located in this stratum. The basal dendrites of Pyramidal neurons are also found here, where they receive input from other Pyramidal cells, septal fibers, and commissural fibers from the contralateral hippocampus (usually recurrent connections, especially in CA3 and CA2.) In rodents the two hippocampi are highly connected, but in primates this commissural connection is much sparser [3].

Stratum pyramidale: It contains the cell bodies of the Pyramidal neurons, which are the principal excitatory neurons of the hippocampus (**Figure 2**). This stratum tends to be one of the more visible strata to the naked eye. In region CA3, this stratum contains synapses from the mossy fibers that course through stratum lucidum. This stratum also contains the cell bodies of many interneurons, including axo-axonic cells, bistratified cells, and radial trilaminar cells [3].

Stratum Luciderm: It is one of the thinnest strata in the hippocampus and only found in the CA3 region. Mossy fibers from the dentate gyrus granule cells course through this stratum in CA3, though synapses from these fibers can be found in statim luciderm [3].

Stratum Radiatum: Like stratum oriens, it contains septal and commissural fibers. It also contains Schaffer collateral fibers, which are the projection forward from CA3 to CA1. Some interneurons that can be found in more superficial layers can also be found here, including basket cells, bistratified cells, and radial trilaminar cells [3].

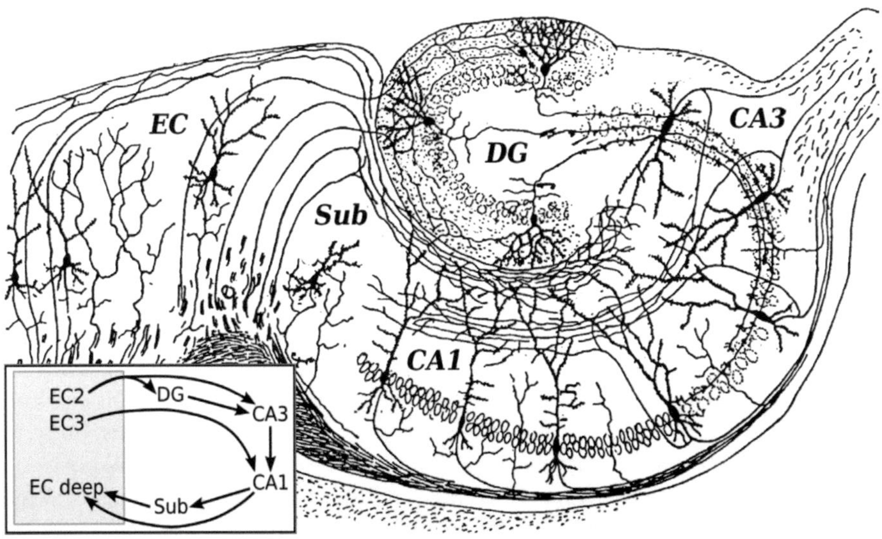

Figure 2.
Layers of the hippocampus.

Stratum Lacunosum: It is a thin stratum that too contains Schaffer collateral fibers, but it also contains perforant path fibers from the superficial layers of entorhinal cortex. Due to its small size, it is often grouped together with stratum moleculare into a single stratum called stratum lacunosum-moleculare [3].

Stratum Moleculare: It is the most superficial stratum in the hippocampus. Here, the perforant path fibers form synapses onto the distal, apical dendrites of Pyramidal cells [3].

Hippocampal Sulcus: Hippocampal Sulcus or fissure is a cell-free region that separates the CA1 field from the dentate gyrus. Because the phase of recorded theta rhythm varies systematically through the strata, the sulcus is often used as a fixed reference point for recording Electroencephalogram (EEG) as it is easily identifiable [3].

Dentate Gyrus: The dentate gyrus is composed of a similar series of strata.

The Polymorphic Layer: It is the most superficial layer of the dentate gyrus and is often considered a separate subfield (as the hilus). This layer contains many interneurons, and the axons of the dentate granule cells pass through this stratum on the way to CA3 [4].

Stratum Granulosum: It contains the cell bodies of the dentate granule cells.

Stratum Moleculare, Inner Third: It is where both commissural fibers from the contralateral dentate gyrus run and form synapse as well as where inputs from the medial septum terminate, both on the proximal dendrites of the granule cells [4].

Stratum Moleculare: External two-thirds is the deepest of the strata, sitting just superficial to the hippocampal sulcus across from stratum moleculare in the CA fields. The perforant path fibers run through this strata and making excitatory synapses onto the distal apical dendrites of granule cells [4].

3. The hippocampal system

The role of the hippocampus in relational memory is in binding together multiple inputs to create and allow for the storage of representations of the associations among the constituent elements of scenes and events [7]. This function ultimately results in the storage of long-term memory in widespread cortical regions. The hippocampus communicates with widespread regions of cortex through a group of highly interconnected brain regions in the medial temporal lobe. Therefore, aberrant activation of the hippocampus would affect perceptual cortical regions, especially those showing high functional connectivity with the hippocampal system.

The hippocampal system consists of the dentate gyrus, cornu ammonis (CA) fields, and the subiculum. The dentate gyrus is an input region, which receives input from the entorhinal cortex. The cornu ammonis (CA) fields of the hippocampus consist of pyramidal cells and are usually subdivided into four regions (CA1–CA4). The area that is often referred to as the parahippocampal gyrus in humans actually consists of several subregions.

The dorsal part of the parahippocampal gyrus (inferior to the hippocampal fissure), throughout its extent, is called the subiculum [8]. The entorhinal cortex provides the major input to the hippocampus and also receives output from the CA1 layer via the subiculum [9]. The entorhinal cortex provides input to the hippocampus through two pathways, one projecting to the dentate gyrus and CA3 fields and the other to CA1 and the subiculum. The subiculum then sends a major input back to the entorhinal cortex.

The entorhinal cortex has its main reciprocal connections with the perirhinal and parahippocampal cortices. Hence, the hippocampus communicates with widespread cortical areas through the entorhinal, perirhinal, and parahippocampal cortices [10]. The hippocampus also contains fiber pathways that run longitudinally throughout its extent [11]. This would allow for the excitation between disparate portions of the hippocampal formation. The entorhinal cortex extends from the amygdala (anteriorly) to approximately 10 mm posterior to the most anterior aspect of the hippocampal fissure.

It had been reported that the hippocampus allows for the consolidation of memory in other cortical regions and that this process proceeds over a number of years. Hence, long-term memory is not actually stored in the hippocampus or, if it is, the hippocampal representation is not necessary for the retrieval of long-term memories after a period of several years. The conceptualization of memory and perception as separate is at odds with this formulation of hippocampal function. Information flows from the cortex to the hippocampus and back out to cortex; this is how memories are formed. In other words, memories are formed via the interplay over time between perceptual regions, including higher-order association cortex and the hippocampus [12, 13].

The physiology of the hippocampus is unique and endows the region with a high level of plasticity that is important for learning and memory; this property also has important implications for neurodegenerative disorders. Neurogenesis also occurs in the hippocampus; hence, it undergoes changes throughout the lifespan. The hippocampus, particularly, the CA1 layer (output layer), one subfield, has the highest concentration of N-methyl-D-aspartate (NMDA) receptors in the brain [14]. NMDA receptors are a type of glutamate receptor whose activity underlies long-term potentiation (LTP), a process that may underlie learning and memory [15].

Neuronal activity reflects the fact that higher-order representational regions from widespread brain regions converge within the hippocampal system. Single unit (neuronal) activity has been shown to be related to a wide variety of stimuli within various tasks or contexts in humans and other animals, including words, pictures locations, odors, and sounds [16]. Unit recording studies also show that although hippocampal system function may not be necessary for the maintenance of short-term memory or working memory, it is active during these types of memory tasks. In fact, if the contents of working memory cannot be actively rehearsed or if this process is interrupted, then the hippocampus is needed to hold the memory, even at short time periods [17]. Hence, trace conditioning is often affected by hippocampal lesions (where there is a temporal gap between the stimuli used), whereas other types of conditioning are intact [18].

An examination of the connectivity of the hippocampal system along with data from single unit recording in the hippocampus necessitates the view that hippocampus is active during much of daily life. For example, the place cells that are recorded in the hippocampus are active regardless of whether or not the memory of spatial location is required at the moment of recording a result also seen for nonspatial stimuli [19]. The online or continuously active role of the hippocampus has recently been formally investigated in recent human neuroimaging studies [20].

The hippocampus may create memory using automatic, obligatory, and ongoing binding operations. Relational memory theory posits that hippocampal-dependent relational processing permits the integration and comparison of discrete experiences and items. In this manner, the hippocampus facilitates the maintenance and integration of the contents of consciousness (consciously perceived stimuli) with representations that are just outside the current contents of consciousness [21].

In this way, the hippocampus could allow for the near-simultaneous activation of representations in cortex that were originally processed with a longer time gap between them. This type of simple mechanism could allow for the association of perceptual stimuli with internally activated memories or representations, resulting in the integration of incoming stimuli with existing cortical associative networks.

The unique physiology of the hippocampus and high concentrations of NMDA receptors allows for relatively high levels of plasticity that are needed for declarative learning and memory. However, this property also confers a unique vulnerability; NMDA receptor abnormalities have also been proposed to play a major role in neurodegenerative disorders [22].

In addition to being the most frequent cite of damage after anoxia or ischemia, the hippocampal system (along with the adjacent amygdala) is the most frequent cite of epileptic foci [23]. The sensitivity of the hippocampus to insult may play a role in the development of epilepsy following traumatic brain injury [24]. It has been reported that hippocampus contains the highest concentration of glucocorticoid (stress hormone) receptors in the brain. These stress hormones can regulate LTP and may increase the likelihood of excitotoxic cell death with prolonged exposure (**Figure 3**) [25].

3.1 The involvement of hippocampus in memory

The hippocampus, for example, is essential for memory function, particularly the transference from short- to long-term memory, control of spatial memory and behavior. The hippocampus is one of the few areas of the brain capable actually growing new neurons, although this ability is impaired by stress-related glucocorticoids. The amygdala also performs a primary role in the processing and memory of emotional reactions and social and sexual behavior, as well as regulating the sense of smell [26].

Another subcortical system (inside the cerebral cortex) which is essential to memory function is the basal ganglia system, particularly the striatum (or neostriatum) which is important in the formation and retrieval of procedural memory. The hippocampal region has been linked to memory function since patient H.M. was first described [27], the hippocampus itself has only recently been identified as a critical structure. Neuropathological

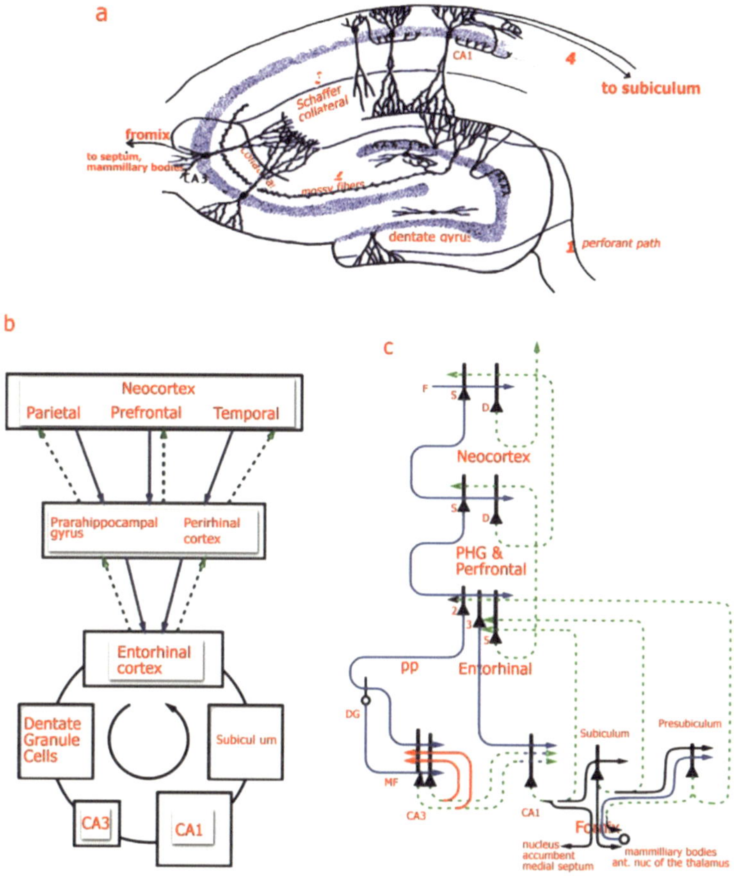

Figure 3.
Representation of connection within the hippocampus. Source: (rolls, 2017).

findings from a patient with permanent circumscribed memory impairment following global ischemia revealed bilateral lesion involving the entire CA1 field of the hippocampus [28]. As a result of this, damage to the hippocampus itself is sufficient to produce clinically significant and long-lasting memory impairment. Furthermore, it has been reported that high-resolution Magnetic Resonance Imaging (MRI) studies of patients with circumscribed memory impairment revealed that the hippocampal formation was reduced in size [29, 30]. The largest area of activation in the memory recall task was in the posterior medial temporal lobe (PMTL) in the region of the hippocampus and the parahippocampal gyrus. There is no activation at the amygdala. Indeed, even incomplete damage to the hippocampus is sufficient to impair memory. The two structures most frequently implicated have been the mammillary nuclei (MN) and the mediodorsal thalamic nucleus (MD) [31, 32]. Idea that damage to the MN impairs memory originated in the finding that the MN are consistently damaged in alcoholic Korsakoff's syndrome.

3.2 Functions of hippocampus

Three phases of memory include (1) registration, (2) storage, and (3) retrieval of information. The hippocampus, parahippocampal region of the medial temporal lobe, and the neocortical association area have been shown through autopsy and imaging studies to be essential for memory processing. Impairment of short-term memory leading up to an inability to form new memories occurs when there is bilateral damage to the above-mentioned regions [33]. The hippocampus is closely associated with the amygdala, hypothalamus, and mammillary bodies such that any stimulation of the nearby parts also marginally stimulates the hippocampus. There are also high outgoing signals from the hippocampus, especially through the fornix into the anterior thalamus, hypothalamus, and greater limbic system. The hippocampus is also very hyperexcitable, meaning it can sustain weak electrical stimulus into a long, sustained stimulation that helps in encoding memory from olfaction, visual, auditory, and tactile senses.

In lower animals, the hippocampus helps them determine if they will eat certain foods, based on olfactory discernment, avoid danger, respond to sexual signals through pheromones, or react to life and death decisions. The hippocampus is a site for decision-making and committing information to memory for future safety uses. Thus, it has a mechanism to convert short-term memory into long-term memory, consolidating the verbal and symbolic thinking into information that can be accessed when needed for decision-making [33].

3.3 Acetylcholine in the hippocampus

It has been shown that cholinergic neurons in the medial septum regulate hippocampal circuits. Optogenetic stimulation of cholinergic neurons in the medial septum area not only causes changes in the firing activity of hippocampal neurons but also modulates theta-band oscillations in the hippocampus in vivo [34]. Experimental and computer modeling studies have shown that Ach specifically inhibits intrinsic pathways, which are part of the memory consolidation circuits, while facilitating afferent projections, which are part of the encoding pathway [35]. Acetylcholine inhibits the recurrent pathway in the CA3 region via the activation of muscarinic ACh receptors in interneurons [36]. This ensures that the circuits that carry extrinsic information are preferentially activated, while the intrinsic projections are toned down [37]. In the hippocampal CA1 region, ACh is known to potentiate the Schaffer collateral pathway, via the activation of α7 or non-α7 nicotinic ACh receptors located in pyramidal neurons and GABAergic interneurons [38]. However, these results are controversial. For example, other studies have shown that the Schaffer collateral pathway is instead inhibited by ACh [39]. One explanation for this discrepancy is that the effect of ACh on synaptic plasticity is timing-dependent. It has been reported that cholinergic input can cause either long-term potentiation or short-term depression, depending on the timing of cholinergic input relative to glutamatergic input to the CA1 [40, 41]. In addition, the effect of ACh may vary depending on which cholinergic receptor subtype is activated in different conditions. In the dentate gyrus, ACh has been shown to increase long-term potentiation via activation of nicotinic and muscarinic receptors [42, 43]. In addition, septal cholinergic projections have been shown to activate astrocytes to modulate dentate granule cells [44].

3.4 Complexity of acetylcholine (ach) modulation on cognitive function

a. ACh's modulation of cognitive function is selective to hippocampus-dependent memory, specifically affecting spatial but not procedural memory.

b. The method of cholinergic modulation can result in different outcomes. For example, optogenetic manipulation of cholinergic neurons is more physiological but is not as selective as antagonism of ACh receptors (AChRs).

c. ACh has differential effects on memory encoding and consolidation, favoring the pathways involved in encoding.

d. ACh can exert different responses depending on which receptor subtypes ACh activates, each of which has distinct desensitization characteristics.

3.5 Diseases of the hippocampus

An understanding of the location and functions of the hippocampus will give us a better idea about the associated diseases. The hippocampus is the part of the brain located in the inner fold of the bottom middle section of the brain. It is a part of the limbic system responsible for the management of feeling and reacting. The main function is for human learning and memory. It is responsible for the retrieval of the main types of memories namely declarative and spatial memory, and there are also short and long term memories.

Hippocampus though known to be important in learning and memory but also important in:

a. Spatial navigation: the process by which organisms use multiple cue sources such as path integration, magnetic cues, landmarks, and beacons to determine the route to a goal and then travel that route. The hippocampus do this by functioning like an internal GPS helping to figure out where we are, have we been here before, and where we can go next [45].

b. Emotional behavior: though emotional behavior is mainly regulated by the amygdala, the hippocampus and amygdala both have reciprocal connections, which can hereby influence each other (latter affects emotions more than former) [46].

c. Regulation of hypothalamic functions: due to the fact that the hippocampus has projections to hypothalamus, hereby affecting the release of adrenocorticotropic hormones. The more reason why patients with atrophied hippocampus have increasing levels of cortisol [47].

Declarative memories are related to facts and events, while spatial memories involve pathways or routes which are stored in the right hippocampus, and short-term memories are converted into long-term memories in the hippocampus and stored in other parts of the brain. Since the hippocampus is a plastic and vulnerable structure that gets damaged by a variety of stimuli resulting in a variety of neurological and psychiatric disorders which produce changes ranging from molecules to morphology. The following are some of the common conditions in which atrophy of human hippocampus has been reported including long-term exposure to high levels of stress [48].

4. Alzheimer's disease

Atrophy of hippocampal region in brains is one of the most consistent characteristics of Alzheimer's disease (AD). It is regarded as the earliest brain region and the most severely affected. A popular hypothesis called "hippocampo-cortical-dissociation" has proposed that early damage to hippocampus causes "dissociation" between hippocampus and cerebral cortex, leading to failure of registration of information emanating from hippocampus. Some amount of hippocampal atrophy is seen in all patients with AD [48]. A number of neurotransmitter alterations can also occur in brains of AD such as noradrenergic, serotonergic, and glutaminergic regions and corresponding loss of neuron in the hippocampal region.

4.1 Depression and stress

Ever since the biological basis of depression is getting revealed, there have been accumulating evidence that prolonged depression can lead to volume loss of hippocampus. With the duration of depression has been correlated with severity of hippocampal atrophy. Evidence suggests that atrophy produced may be permanent and persist long even though depression has undergone remission. It has been hypothesized that it could result to affective disturbance seen in depression. It is believed that this could be as a result of prolonged stress generated as a result of depression. Retraction of cell volume and/or suppression of hippocampal neurogenesis could be responsible in this case [49].

4.2 Schizophrenia

There is a reason to believe that disturbance in hippocampus is responsible for the production of psychotic symptoms in schizophrenia. Hippocampal volume reduction is one of the most consistent findings found in MRI of schizophrenic patients. Though functional and biochemical abnormalities have also been identified initially the pathophysiology of schizophrenia mainly focusing on prefrontal cortex, now hippocampus is being considered for last 20 years or so. There is now a compelling data to suggest that there are anatomical and functional aberrations as a result of neuronal disturbances in hippocampus of schizophrenic patients. Evidence gathered from MRI, Positron Emission Tomography (PET), and Magnetic Resonance Spectroscopy (MRS) studies of disturbances within hippocampus of schizophrenia. Volume reduction in hippocampus of schizophrenia is modest and not as marked as that seen in AD. Still, biochemical and functional disturbances provide a reliable evidence of involvement of hippocampus in pathophysiology of schizophrenia [49].

4.3 Epilepsy

Up to 50–75% of patients with epilepsy may have hippocampal sclerosis upon postmortem analysis, in case they died and had medically refractory temporal lobe epilepsy. It is, however, not clearly known if epilepsy is generated as a result of hippocampal sclerosis or repeated seizures damage hippocampus. Therefore, there is not much clarity whether hippocampal atrophy is a cause or consequence of recurrent seizures [50, 51] Mechanism of hippocampal sclerosis in epilepsy that has been reported might be related to the development of uncontrolled local hippocampal inflammation and blood–brain barrier damage [51] Recurrent seizures leading to cytoskeletal abnormalities, neurotransmitter alterations, and hypoxia may be additional associated factors. Developing hippocampus may be more susceptible to damage compared to mature one [50]. Recent evidence also suggests that hippocampal sclerosis in epilepsy may be an acquired process with accompanying re-organizational dysplasia and an extension of mesial temporal sclerosis rather than a separate pathological entity [52]. Indicating significant progress is being made in understanding relationship between hippocampal sclerosis and epilepsy. It is believed that hippocampus has an inhibitory effect on seizure threshold (i.e. it keeps it elevated). Once it gets damaged, then seizures become more intractable. Surgical resections of the hippocampus have been suggested as the most successful treatment for medication-refractory medial temporal lobe epilepsy due to hippocampal sclerosis [53]. It was also discovered that patients with AD with seizures had hippocampal atrophy, and cause of hippocampal atrophy in epilepsies is not known, but autoimmunity has been proposed as one of the mechanisms [54, 55].

4.4 Food that enhances the hippocampus

Brain foods as well as hippocampus are those that are rich in antioxidants, healthy fats, vitamins, and minerals. They provide your brain with energy and aid in protecting brain cells, which helps ward off development of brain diseases such as dementia. There is no single brain food can ensure a sharp brain as age declines. Nutritionists emphasize that the most important strategy is to follow a healthy dietary pattern that includes a lot of fruits, vegetables, legumes, and whole grains. Diet that is rich in omega-3 fatty acids is garnering appreciation for supporting cognitive processes in humans [56] and upregulating genes that are important for maintaining synaptic function and plasticity in rodents [57].

Fish (salmon), flax seeds, krill, chia, kiwi fruit, butternuts, and walnuts are sources of omega-3 fatty acids (e.g. docosahexaenoic acid). They have ameliorating effect of cognitive decline in the elderly [58], basis for treatment in patients with mood disorders [59], improvement of cognition in traumatic brain injury in rodents [60], and amelioration of cognitive decay in mouse model of Alzheimer's disease [61]. Butter, ghee, suet, lard, coconut oil, cottonseed oil, palm kernel oil, dairy products (cream and cheese), and meat which are sources of saturated fat promote cognitive decline in adult rodents [62], aggravation of cognitive impairment after brain trauma in rodents [60], and exacerbation of cognitive decline in aging humans.

5. Conclusions

Hippocampus is an extension of cerebral cortex situated deep into temporal lobe. It is a vulnerable and plastic structure. It gets damaged by a variety of stimuli and hence is important clinically both diagnostically and therapeutically. Currently, it is one of the markers of cognitive decline and diagnosis of AD. It is also a prognostic marker in research setting. Drugs that are able to cause slow-down of atrophy or reversal are actively being sought. These could then potentially have disease-modifying effects.

Author details

Lawrence Adedayo[1*], Gideon Ojo[2], Samuel Umanah[3], Gideon Aitokhuehi[4], Ileri-Oluwa Emmanuel[5] and Olubayode Bamidele[1]

1 Neurophysiology Unit, Physiology Program, Bowen University Iwo, Nigeria

2 Anatomy Program, Bowen University Iwo, Nigeria

3 Department of Anatomy, University of Uyo, Nigeria

4 Department of Physiology, University of Medical Sciences, Ondo City, Nigeria

5 Biomedicinal Research Center, Forestry Research Institute of Nigeria, Nigeria

*Address all correspondence to: delawbaba@gmail.com

References

[1] Ananya M. What is the hippocampus? News-Medical. 2019. Available from: https://www.news-medical.net/health/Hippocampus-What is the hippocampus.aspx [Accessed: January 6, 2023].

[2] Carr MF, Jadhav SP, Frank LM. Hippocampal replay in the awake state: A potential substrate for memory consolidation and retrieval. Nature Neuroscience. 2011;**14**:147-153

[3] Gluck M, Mercado E, Myers C. Learning and Memory from Brain to Behavior. Second ed. New York: Kevin Feyen; 2014. p. 416

[4] Daugherty AM, Bender AR, Raz N, Ofen N. Age differences in hippocampal subfield volumes from childhood to late adulthood. Hippocampus. 2016;**26**(2):220-228

[5] Wang EW, Du G, Lewis MM, Lee EY, De Jesus S, Kanekar S, et al. Multimodal MRI evaluation of parkinsonian limbic pathologies. Neurobiological Aging. 2019;**76**:194-200

[6] Spallazzi M, Dobisch L, Becke A, Berron D, Stucht D, Oeltze-Jafra S, et al. Hippocampal vascularization patterns: A high-resolution 7 tesla time-of-flight magnetic resonance angiography study. Neuroimage Clinicals. 2019;**21**:101609

[7] Cohen NJ, Ryan J, Hunt C, Romine L, Wszalek T, Nash C. Hippocampal system and declarative (relational) memory: Summarizing the data from functional neuroimaging studies. Hippocampus. 1999;**9**:83-98

[8] Suzuki WA, Amaral DG. Perirhinal and parahippocampal cortices of the macaque monkey: Cortical afferents. Journal Comparative Neurology. 1994;**350**:497-533

[9] Witter MP, Amaral DG. Entorhinal cortex of the monkey: V. Projections to the dentate gyrus, hippocampus, and subicular complex. Journal Comparative Neurology. 1991;**307**:437-459

[10] Lavenex P, Suzuki WA, Amaral DG. Perirhinal and parahippocampal cortices of the macaque monkey: Projections to the neocortex. Journal of Neurology. 2002;**447**:394-420

[11] Sloviter RS, Lomo T. Updating the lamellar hypothesis of hippocampal organization. Frontiers in Neural Circuits. 2012;**6**:102

[12] Rempel-Clower NL, Zola SM, Squire LR, Amaral DG. Three cases of enduring memory impairment after bilateral damage limited to the hippocampal formation. The Journal of Neuroscience. 1996;**16**:5233-5255

[13] Reed JM, Squire LR. Retrograde amnesia for facts and events: Findings from four new cases. The Journal of Neuroscience. 1998;**18**:3943-3954

[14] Cotman CW, Monaghan DT. Anatomical organization of excitatory amino acid receptors and their properties. Advances Experimental Medicine. 1986;**203**:237-252

[15] Bliss TV, Lomo T. Long-lasting potentiation of synaptic transmission in the dentate area of the anaesthetized rabbit following stimulation of the perforant path. The Journal of Physiology. 1973;**232**:331-356

[16] Quiroga RQ, Kreiman G, Koch C, Fried I. Sparse but not ‗grandmother-cell‘coding in the medial temporal lobe. Trends in Cognitive Sciences. 2008;**12**:87-91

[17] Hannula DE, Tranel D, Cohen NJ. The long and the short of it: Relational memory impairments in amnesia, Even at short lags. Journal of Neuroscience. 2006;**26**:8352-8359

[18] Weiss C, Bouwmeester H, Power JM, Disterhoft JF. Hippocampal lesions prevent trace eyeblink conditioning in the freely moving rat. Behavioural Brain Research. 1999;**99**:123-132

[19] Thompson LT, Best PJ. Long-term stability of the place-field activity of single units recorded from the dorsal hippocampus of freely behaving rats. Brain Research. 1990;**509**:299-308

[20] Warren DE, Duff MC, Jensen U, Tranel D, Cohen NJ. Hiding in plain view: Lesions of the medial temporal lobe impair online representation. Hippocampus. 2012;**22**:1577-1588

[21] Eichenbaum H. Hippocampus: Cognitive processes and neural representations that underlie declarative memory. Neuron. 2004;**44**:109-120

[22] Javitt DC. When doors of perception close: Bottom-up models of disrupted cognition in schizophrenia. Annual Review of Clinical Psychology. 2009;**5**:249-275

[23] Wiebe S. Epidemiology of temporal lobe epilepsy. Behavioral Science. 2013;**3**:311

[24] Griesemer D, Mautes AM. Closed head injury causes hyperexcitability in rat hippocampal CA1 but not in CA3 pyramidal cells. Journal of Neurotrauma. 2007;**24**:1823-1832

[25] Sandi C. Glucocorticoids act on glutamatergic pathways to affect memory processes. Trends Neuroscience. 2011;**34**:165-176

[26] Mastin L. Parts of the Brain. The Human Memory. 2017;**4**:50-62

[27] Scoville WB, Milner B. Loss of recent memory after bilateral hippocampal lesions. Journal of Neurology, Neurosurgery, and Psychiatry. 1957;**20**:1151

[28] Victor M, Agamanolis D. Amnesia due to lesions confined to the hippocampus: A clinical-pathologic study. Journal of Cognitive Neuroscience. 1990;**2**:246-257

[29] Squire L, Zola-Morgan S. The medial temporal lobe memory system. Science. 1991;**253**:1380-1386

[30] Press G, Amaral D, Squire L. Hippocampal abnormalities in amnesic patients revealed by high resolution magnetic resonance imaging. In: Nature. Vol. 341. New York: Press; 1989. pp. 54-67

[31] Markowitsch HJ. Diencephalic amnesia: a reorientation towards tracts? Brain Research Reviews. 1988;**13**:351-370

[32] Victor M, Adams R, Collins G. The Wernicke-Korsakoff Syndrome and Related Neurological Disorders due to Alcoholism and malnutrition. 2nd ed. Philadelphia. Davis; 1989

[33] Kizilirmak JM, Schott BH, Thuerich H, Sweeney-Reed CM, Richter A, Folta-Schoofs K, et al. Learning of novel semantic relationships via sudden comprehension is associated with a hippocampus-independent network. Conscious Cognition. 2019;**69**:113-132

[34] Dannenberg H, Pabst M, Braganza O, Schoch S, Niediek J, Bayraktar M, et al. Synergy of direct and indirect cholinergic septo-hippocampal pathways coordinates firing in hippocampal networks. The Journal of Neuroscience. 2015;**35**:8394-8410

[35] Newman EL, Gupta K, Climer JR, Monaghan CK, Hasselmo ME.

Cholinergic modulation of cognitive processing: insights drawn from computational models. Frontiers in Behavioral Neuroscience. 2012;**6**:24

[36] Hasselmo ME, Schnell E, Barkai E. Dynamics of learning and recall at excitatory recurrent synapses and cholinergic modulation in rat hippocampal region CA3. Journal of Neuroscience. 1995;**15**:5249-5262

[37] Hasselmo ME. Neuromodulation: acetylcholine and memory consolidation. Trends in Cognitive Sciences. 1999;**3**:351-359

[38] Leao RN, Mikulovic S, Leao KE, et al. OLM interneurons differentially modulate CA3 and entorhinal. Nature Neuroscience. 2012;**15**:1524-1530

[39] Mans RA, Warmus BA, Smith CC, McMahon LL. An acetylcholinesterase inhibitor, eserine, induces long-term depression at CA3-CA1 synapses in the hippocampus of adult rats. Journal of Neurophysiology. 2014;**112**:2388-2397

[40] Ji D, Lape R, Dani JA. Timing and location of nicotinic activity enhances or depresses hippocampal synaptic plasticity. Neuron. 2001;**31**:131-141

[41] Gu Z, Yakel JL. Timing-dependent septal cholinergic induction of dynamic hippocampal synaptic plasticity. Neuron. 2011;**71**:155-165

[42] Luo L, Chen WH, Wang M, Zhu DM, She JQ, Ruan DY. Modulation of long-term potentiation by individual subtypes of muscarinic acetylcholine receptor in rat dentate gyrus. Hippocampus. 2008;**18**:985-995

[43] Welsby PJ, Rowan MJ, Anwyl R. Intracellular mechanisms underlying the nicotinic enhancement of LTP in the rat dentate gyrus. European Journal of Neuorscience. 2009;**29**:65-75

[44] Pabst M, Braganza O, Dannenberg H, et al. Astrocyte intermediaries of septal cholinergic modulation in the hippocampus. Neuron. 2016;**90**:853-865

[45] Stella F, Cerasti E, Si B, Jezek K, Treves A. Self-organization of multiple spatial and context memories in the hippocampus. Neuroscience and Biobehavioral Reviews. 2011;**36**(7):1609-1625

[46] Toyoda H, Li XY, Wu LJ, Zhao MG, Descalzi G, Chen T, et al. Interplay of amygdala and cingulate plasticity in emotional fear. Neural Plasticity. 2011;**2011**:813749

[47] Koehl M, Abrous DN. A new chapter in the field of memory: Adult hippocampal neurogenesis. The European Journal of Neuroscience. 2011;**33**:1101-1114

[48] Dhikav V, Anand K. Potential predictors of hippocampal atrophy in Alzheimer's disease. Drugs & Aging. 2011;**28**:1-11

[49] Bast T. The hippocampal learning-behavior translation and the functional significance of hippocampal dysfunction in schizophrenia. Current Opinion in Neurobiology. 2011;**21**:492-501

[50] Cendes F. Febrile seizures and mesial temporal sclerosis. Current Opinion in Neurology. 2004;**17**:161-164

[51] Yang T, Zhou D, Stefan H. Why mesial temporal lobe epilepsy with hippocampal sclerosis is progressive: Uncontrolled inflammation drives disease progression? Journal of the Neurological Sciences. 2010;**296**:1-6

[52] Thom M, Eriksson S, Martinian L, Caboclo LO, McEvoy AW, Duncan JS, et al. Temporal lobe sclerosis associated with hippocampal sclerosis in temporal

lobe epilepsy: Neuropathological features. Journal of Neuropathology Experimental Neurology. 2009;**68**:928-938

[53] Bonilha L, Martz GU, Glazier SS, Edwards JC. Subtypes of medial temporal lobe epilepsy: Influence on temporal lobectomy outcomes? Epilepsia. 2012;**53**:1-6

[54] Dhikav V, Anand KS. Is hippocampal atrophy a future drug target? Medical Hypotheses. 2007;**68**:1300-1306

[55] Dhikav V, Anand KS. Glucocorticoids may initiate Alzheimer's disease: A potential therapeutic role for mifepristone (RU-486). Medical Hypotheses. 2007;**68**:1088-1092

[56] McCann JC, Ames BN. Is docosahexaenoic acid, an n-3 long-chain polyunsaturated fatty acid, required for development of normal brain function? An overview of evidence from cognitive and behavioral tests in humans and animals. The American Journal of Clinical Nutrition. 2005;**82**:281-295

[57] Wu A, Ying Z, Gomez-Pinilla F. Omega-3 fatty acids supplementation restores mechanisms that maintain brain homeostasis in traumatic brain injury. Journal of Neurotrauma. 2007;**24**:1587

[58] Van Gelder BM, Tijhuis M, Kalmijn S, Kromhout D. Fish consumption, n-3 fatty acids, and subsequent 5-y cognitive decline in elderly men: The Zutphen elderly study. The American Journal of Clinical Nutrition. 2007;**85**:1142-1147

[59] Freeman MP et al. Omega-3 fatty acids: Evidence basis for treatment and future research in psychiatry. The Journal of Clinical Psychiatry. 2006;**67**:1954-1967

[60] Wu A, Molteni R, Ying Z, Gomez-Pinilla F. A saturated-fat diet aggravates the outcome of traumatic brain injury on hippocampal plasticity and cognitive function by reducing brain-derived neurotrophic factor. Neuroscience. 2003;**119**:365-375

[61] Hashimoto M et al. Chronic administration of docosahexaenoic acid ameliorates the impairment of spatial cognition learning ability in amyloid β-infused rats. The Journal of Nutrition. 2005;**135**:549-555

[62] Molteni R, Barnard JR, Ying Z, Roberts CK, Gomez-Pinilla F. A high-fat, refined sugar diet reduces hippocampal brain-derived neurotrophic factor, neuronal plasticity, and learning. Neuroscience. 2002;**112**:803-814

Chapter 3

Perspective Chapter: Role of the Hippocampal Formation in Navigation from a Simultaneous Location and Mapping Perspective

André Pedro, Jânio Monteiro and António João Silva

Abstract

The research of the brain has led to many questions, with most of them still not having a definitive answer. One of those questions is about how the brain acts when we navigate a new space. Inside the Temporal Lobe's Hippocampal structure, specific types of neurons and neuronal structures are responsible to identify spatial elements. To recognize spaces, these cells require data, which is obtained from the subject's senses. It is important to understand how these features are captured, processed, encoded and how the Hippocampus, and its neighboring elements, use the information to help in the navigation and mapping of a place. A specific type of neurons seems to support an animals location and spatial mapping, on other areas of research, discrete global grid systems are used to increase the independence of the autonomous vehicles, allowing the indexing of assets across the globe by partitioning the earth into grids that take into account the heterogeneity of the scales of the associated geospatial data. In this context, the main objective of this chapter is to make an analysis about the biological and technical aspects of navigation by establishing a bridge between the Hippocampus and Simultaneous Localization and Mapping (SLAM) methods.

Keywords: hippocampus, entorhinal cortex, SLAM, Grid Cells, robotics, spatial navigation, sensory processing

1. Introduction

The brain is a complex structure that somehow is able to make associations between different inputs from different areas. The information used for these associations results from the body senses, including the Visual, Olfactory, Auditory, Gustatory, Somatosensory and Self-Motion. This process is done without labeling the data entries. It is believed that most of these associations occur in the Hippocampus. The Hippocampus is a Temporal Lobe element and its functions are related to the creation of new memories [1], control of emotions [2] and spatial navigation [3]. There are several conceptual models about the functioning of the Hippocampus, one of them is the memory index theory where the Hippocampus plays the role in creating an index of neocortical pattern associations related to a context or situation [4]. Since the discovery of the Place Cells in the Hippocampus and their connection of spatial processing elements of the Entorhinal Cortex (EC), it is believed that the Hippocampus can identify places and the contextual situations associated with those areas [5]. The information about the state of the senses enters the Hippocampus through the EC, that in turn receives inputs directly from the senses or through supporting elements like the Parahippocampal gyrus and Peririhinal Cortex [5]. The EC area is divided into two different parts, the Lateral Entorhinal Cortex (LEC), that is responsible for the handling of data not related with spatial navigation, and the Medial Entorhinal Cortex (MEC). The Grid Cells are a set of cells located in the MEC, with the assumed ability to identify positions inside a spatial environment. Such behavior is possible due to its hexagonal firing pattern [5]. Also, according with [6], the Grid Cells are organized in a multi layer format. Such organization allows those cells to give the necessary precision in the determination of a spatial position. The information produced by the Grid Cells is sent to the Hippocampus, specially affecting the Place Cells, where the data is used to encode and to create representations of the known places, by adapting to new environmental changes with a mechanism called global remapping [7]. Is important to understand how the senses gather information from the environment and how such data is processed, because a coherent understanding of the space depends in the acquisition of such information. The Vision, the Audition and other senses are crucial for the navigational calculations because those elements are used to determine the environmental landmarks, essential in the creation of a mental map [7]. In this scope, the purpose of this book chapter is to describe the work done until now in this area, focusing in how the obtained sensory information can be used for spatial navigation and how the Hippocampal Formation elements are used to solve navigational problems. This book chapter is composed of five sections. The first section, contains an analysis about the importance of navigation and how these questions are handled with the available computational resources, followed by a second and a third sections about how the brain elements gather, interpret, prepare and encode the sensory information used in spatial navigation applications. The fourth section, focuses in the bridge between the knowledge obtained from neuroscience trials and experiments using the available computational tools to simulate/recreate some of the Hippocampal Formation characteristics, based in computer models dedicated to the Hippocampus and to EC. It ends with a section dedicated to the discussion of the extracted scientific information.

2. The importance of navigation

The ability to navigate is an important function, common to animals. Navigation is useful to avoid dangerous areas and to find the necessary resources for survival, including food and resting places [8].

Complementary to their intrinsic abilities for navigation, Humans initially had few, external, tools to solve navigation problems. Most of the times the stars in the sky were the only reliable reference to determine our position [9]. However, such navigation method is fully dependent from the stars or the sun, which can only be seen during the day, or in a clear night sky, not to mentioned the fact that our perception of the stars change duo to the Earth rotation [10]. To answer such questions several instruments were created, like the Geographical Representation Systems, that map the entire world with a relative good precision or physical tools that helped solving possible errors in the calculation of the position. Also, the navigation task can be seen has a pure mathematical problem [11]. The Simultaneous Localization and Mapping

(SLAM) approach, has the challenge to map unknown environments, supported by different types of algorithms for spatial mapping of a space, contributing to the increase of autonomy in virtual and physical agents.

The following sequence of sub-chapters start with an analysis about the past and present of the Geographical Representation Systems, followed by a perspective about the SLAM characteristics, ending with an in depth analysis about the available methods and strategies used to solve navigational problems.

2.1 Geographical representation systems

As mentioned previously, the initial navigation methods made by humans relied on the observation of the stars like the North Star, the Southern Cross, or using Planets like Jupiter or Saturn [12]. Back in that time, there were several charts with the position of the Stars and Planets in the different hours of the day. The stars were a complementary element of a navigation method called dead reckoning. The Dead Reckoning is a method which allows the calculation of a heading without using heading calculation instruments like the compass. It uses the distance, the known positions recorded in a time period and the estimated drift to determine the direction of travel [13].

In the 1500 a.c, a certain type of instruments such as the Quadrant and the Astrolabio were used to get a better position from the stars, and with them it was possible to discover a sea route from Europe to Brazil [14]. Later, the navigational tasks become easier with the help of another instrument called the Gyroscope, that had the function to determine the horizon line [15].

At that time, it became clear the necessity to have a system which could represent with a certain precision all the regions of the globe. The Mercator projection divides the world into a squared grid, and its basis are still used in our days, despite the fact that this representation method was developed in the XVI century [16]. Nowadays, there are new forms to find and represent spatial information. The Global Positioning System (GPS), that uses the trilateration method where the distance between a device to a set of three or more satellites is used to calculate the position, as demonstrated in **Figure 1** [16], and the Web Mercator projection, that allows the implementation of Mercator projection characteristics in digital platforms. Also, the Equal Earth projection is used to represent the geographical surface using the continent area proportions. The combination between a Polynomial Equation with the Robinson projection, are the basis of the method that has an easy implementation and it is a good attempt to represent the Earth in the fairest possible way [18].

Also, there are other types of systems with the capability to represent the surface of the Earth. The Discrete Global Grid Systems (DGGS) are a digital geographic representation form where the projections are formed using geometrical forms, called cells, in a hierarchical structure. The main function of DGGS consists in creating maps of information providing a digital framework used in Geographical Information Systems [19]. The implementations of this approach can be found in Hexagonal Hierarchical Geospatial Indexing System (H3), developed by the Uber Tecnologies Inc., and the Open Equal Area Global Grid (OpenEAGGR) where the cells have the same size, and it has the possibility to convert from latitude and longitude references, to cell elements [20]. Although, the DGGS systems seem to be used in the representation of geographical information, they appear to have difficulties in the aggregation of similar information. According, with [21], the best methods to index spatial data are the ones that use the triangle from has a base, where its edges are bisected, forming sub-triangles, allowing the increase of precision.

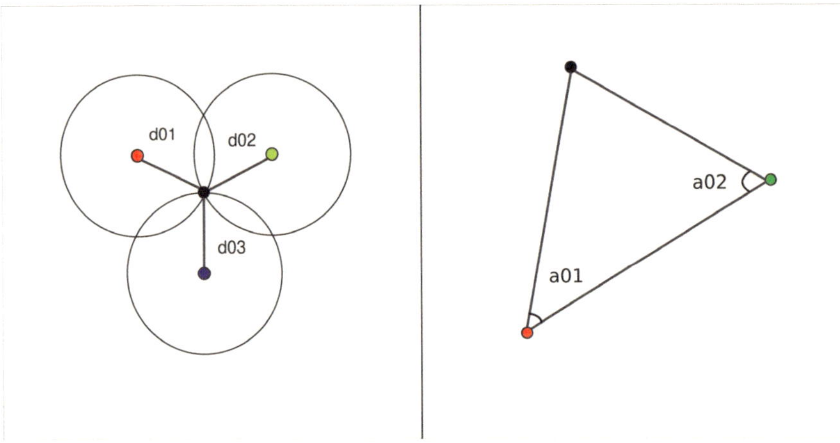

Figure 1.
Demonstration of the difference between trilateration and triangulation, adapted from [17]. The left image is the trilateration method where the distances are used to determine a location. The image on the right is the triangulation method, where the position of the objective (black dot) is determined by the angles a01 and a02 that connect the reference points to the objective.

2.2 Simultaneous localization and mapping

The SLAM problem involves creating a map of an unknown environment while simultaneously determining the location of a mobile agent, virtual or physical, within that environment, without any previous information about the space [11]. The solution to the SLAM problem increases the autonomy of the agents, but it is necessary to take into account the environment of the agent, namely if it is indoor, outdoor, underwater or in the air. At first, the SLAM question was approached using a probabilistic mindset, taking into account that is problem has a chicken and egg situation. The robot needs a position has a reference to start the mapping, but how it is possible to give such position where there is no map references [22, 23].

During the mapping process, the agent extract landmark information from its sensors as long as it stays inside the designated area. Depending of the SLAM method, the mapping procedure can be made with statistical strategies or using Machine Learning procedures, like the Deep Learning Neural Networks, where those structures are used to recognize environmental references [24]. To better characterize the behavior of the SLAM methods, the steps can be divided into four parts: (i) the input of the data, which consists in the extraction of environmental information, (ii) the mapping of place based in the acquired data, (iii) the determination of the location using navigation planning strategies and (iv) the output based on the results obtained from the previous steps [25].

Nowadays, the SLAM methods are more accurate and precise because of the development of new types of sensors like the Lidar, that sends light beams which reflects in the objects creating a good map of the space and the Visual SLAM which captures information from the agent's video resources, making the definition of the landmarks less probabilistic and more dependent of the surrounding space. Also, the SLAM approaches can be applied in 3D spaces, allowing vehicles, or agents, to keep track of their position and to map areas while they are navigating in it [26] and they can be seen as a complement to the existing navigational technologies, like the GPS

[27], where the SLAM can complement the other navigational methods, by correcting the drifts and inaccuracies caused by the navigational instruments [28].

2.2.1 Methods and implementations of SLAM

At first, some SLAM solutions used various types of Kalman Filter methods, fusing the sensory data to determine the position based in the location of the landmarks. However, as mentioned in Section 2.2, with the increase of the computational resources and the appearance of new sensors, it was possible to develop new methods that took into account these new technological improvements.

Some of these methods are the Gmapping method [29], which uses the Partical Filters to estimate the position, or the LagoSLAM [30] that works as a graph-based SLAM, where the cost function is a non-linear and non-convex formula. There are also several models that were applied to three dimensional (3D) environments such as the Loam [31], which uses a 3D Lidar sensor to scan the space and the IMLS-SLAM [32] that has a drift reduction algorithm to optimize the results obtained from the three dimensional Lidar. With the appearance of deep learning methods, it was possible to fuse the Artificial Intelligence neural networks into SLAM methods, as in the VoxelNet [24] that uses 3D networks to help in the extraction of features and the SqueezeSeg [33] which uses Convolutional Neural Networks to perform a real-time data segmentation [23].

In terms of the Visual SLAM algorithms there are some approaches like: (i) the ORB-SLAM [34] that uses the ORB computer vision method [35] to define the environmental landmarks, (ii) the OpenVSLam [36] which can be used with different types of cameras and it has algorithms that handle the features in a sparse form and (iii) the VINS [37, 38] that is a SLAM system, with the ability to work in real-time, which can be used in robots running ROS. The VINS can work in mobile platforms like the iOS and it can receive information from different types of sensors like the GPS, to correct some of the method drifts. Deep Learning approaches are also present in Visual SLAM methods, like in the: (i) DeepSLAM [39] that handles some of the noise present in the gathered images; (ii) the ScanComplete [40] model which calculates the position even with incomplete three dimensional scans and (iii) the DA-RNN [41] that has a Recurrent Neural Network to label the spatial data captured by RGB-D cameras [23].

In all of the possible approaches to solve SLAM, there is one that enhances the brain's ability to navigate through different types of spaces and environments. Biological based SLAM, started to appear when the scientific community have found different types of brain cells, using the Hippocampal Formation elements, with the ability to solve navigational tasks [42]. Some of these methods use special types of neural networks such as the Spiking Neural Networks (SNN) to mimic the brain behaviors in the processing of sensory data to map environments and spaces. The details of these types of models will be further discussed in Section 5.

3. The hippocampus and navigation

As mentioned in Section 1, the Hippocampus is a brain element that is present in humans and other animals. It is located in the Temporal Lobe area and is composed of a left and right Hippocampus. Inside it there are several sub-structures essential to the creation of short-term memories, to the recognition of data patterns and to identify spatial areas [4, 43]. With that in mind, it is important to understand how the different sub-structures handle the information, what are the functions of their neural elements and how they work with the data sent though the EC.

In the following sub-chapters are dedicated to the biological analysis of the sub-structures of the Hippocampal loop, starting with the Dentate Gyrus and ending at the Subiculum, as demonstrated in **Figure 2**, maintaining a constant bridge between the Hippocampus functions with their ability to solve navigational problems.

3.1 Dentate gyrus

The Dentate Gyrus (DG) receives information from the layer II of the LEC and MEC. It is considered a sensory data pattern separator element, that removes the disambiguity obtained from the sensory data [45]. The DG seems to allow the discrimination of elements, like the difference of two objects in the same space [45]. The DG is composed by the Granule Cells and Mossy Cells. They are specialized in the processing of sensory information, to make some associations and to detect misplaced

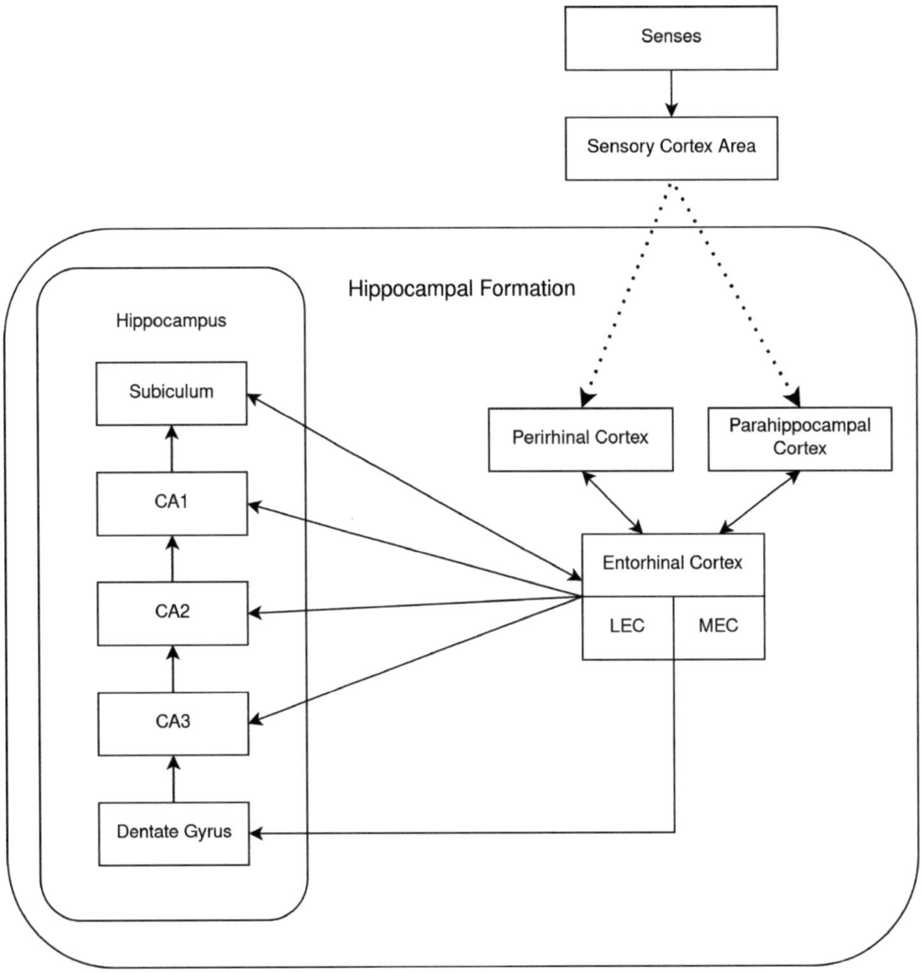

Figure 2.
Simple structure of the hippocampal formation, with the hippocampus closure adapted from [44].

items. It sends the information to the CA3 area. These cells have a low firing rate and they seem to represent the information using a sparse codification [46]. Also, the Mossy Cells act as a data hub which connects the DG to the Mossy Fibers. The Mossy Fibers are connected from the DG into the CA3 area. The Mossy Fibers could act as a conditional detonator or discriminator, because they discharge their energy potential only on special occasions and such discharge inhibits the Cornu Ammontis 3 (CA3) ability to receive information from the layer II of LEC and MEC [47].

3.2 Cornu Ammontis 3

The CA3 region is responsible for the association of episodic events and identification of information patterns, essential in the transformation of sensory events into short-term memories [45]. This area receives input from the DG, via Mossy Fibers, and the layer II of the LEC and MEC. It was discovered that inside the Hippocampus of a rat, the CA3 area has a set of recurrent connections which allows the CA3 to store information required for the pattern completion and memory association process [48]. The CA3 element is mostly composed of Pyramidal Neurons which are the elements responsible for the differentiation and the detection of overlapped memories [49]. The CA3 structure is divided into 3 different parts: the CA3a and CA3b areas seem to be dedicated to the encoding of information, while the CA3c connects to the Mossy Fibers and seems to help in the separation of occurred events [50]. Also, it is believed that the CA3 element could be important to the prediction of future steps and together with DG they could play a role in the correction of path integration calculations [51].

3.3 Cornu Ammontis 2

The Cornu Ammontis 2 (CA2) region acts as a hub between the CA3 and CA1. It is composed of Pyramidal Neurons and they seem to characterize the patterns related to social habits [52]. The CA2 has cells that discharge their energy in different locations of a space. These elements are similar to Place Cells, however their place field, which is the spatial area associated to the neuron, is more suitable to time factor changes, meaning that such elements can change their state in a matter of hours or days. This means that CA2 could create associations between time and place [53].

Also, it appears that CA2 could help in the organization of information in CA1, according with their periods and time characteristics because the information inside the CA1 is more stable to changes [54].

3.4 Cornu Ammontis 1

The Cornu Ammontis 1 (CA1) seems to be responsible for the retrieval of memory elements permitting the recognition of known episodic events, giving the "where" context on those elements [7]. The CA1 component is divided into two different parts, the distal part that is responsible to extract information from the sensory data and the proximal area which isolates the spatial information to allow the recognition of events. The CA1 could discriminate ambiguous information gathered from the MEC and CA3 areas [55]. The data stored in CA1 maintains its stability for long periods of time. Also the CA1 receives projections from the other Hippocampus, via a set of fibers called Hippocampal Comissure Fibers [56].

The CA1 could be used for Navigation purposes, because it has a special group of cells called Place Cells and their task consists in representing a specific spatial area, with the ability to adapt to environmental changes [7]. The area represent by each Place Cells is called place field and when a person or animal passes such area, the correspondent Place Cell discharges its energy [57]. The deepest part of the CA1 area has a link with the Landmark Vector Cells, because their firing is stronger when the agent is in a position of a particular landmark [58].

3.5 Subiculum

The Subiculum is located after the CA areas and it divides into different sections: (1) A section which receives the projections from the CA1 module, (2) a section dedicated to the reinforcement of the learned data and (3) a section that connects to the EC, closing the Hippocampal Loop. Also, the Subiculum can provide information regarding the boundaries of a spatial area [59]. Border Cells fire when a subject is facing a border in their allocentric direction [60]. These cells are present in other areas like the Parasubiculum and the MEC [61].

According to [62], the Subiculum may have the necessary structures to represent the memory of events with more precision and robustness. The data is converted from a sparse format to a denser one and then the processed data is sent to another brain element called Nucleus Accumbens, to decide which step to take at a given time. This could mean that the Hippocampus forms a map of possible decisions valid for a time period.

Also, inside the Subiculum there are cells called Head Direction Cells (HD), that are capable of indicating the direction which a person or animal is facing at a given moment [63]. The HD cells have a dual-axis system used to represent complex spaces and it could process 3D environments due to the possibility to calculate the azimuth using visual data as referential landmarks [64].

4. The connection between hippocampus and senses

The Hippocampus relies on the information gathered from the senses to understand the animal's surroundings. The sensory information is sent to the Hippocampus via the EC. The EC is located in the Temporal Lobe area and it can process the spatial and non-spatial information before it is sent to the Hippocampus. Giving the importance of the EC, it is necessary to understand how the data is encoded in a format, understandable by and Hippocampus and how small changes in the acquired information affect the spatial characterization.

In the following sub-section we start to mention some of the properties of the EC and how they receive and interpret the data from the senses. Then, the focus is directed in how the Vision and Audition are used for navigational purposes. This chapter ends with an analysis about how the brain elements encode the sensory information.

4.1 Entorhinal cortex

The EC area is divided into two distinct parts, the LEC and MEC, each one with different types of neurons.

The LEC receives inputs from the Perirhinal and Parahippocampal corteces. Those elements process data gathered from the olfactory and somatosensory senses [65]. Also, the LEC appears to encode temporal information across different periods of time [66], using a set of cells called Ramping Cells. They act as a supporting element in the acquisition of a time reference because they discharge the energy in a periodic format. The data is further sent to the Hippocampus to a group of cells called Time Cells and they useful for organizing the memory according with their time reference [67].

The MEC is divided into 6 different layers, each one with specific types of neurons and neuronal structures. The first layer is related to the transfer of information between the second and third layers. The elements responsible for such behavior are the Chandelier Cells because they regulate the inhibition response to the Pyramidal Neurons [68]. The second layer is composed of Stellate Cells that support Grid Cells with the ability to transmit the necessary information for the path-integration calculations [69]. The third layer is a mixture of Pyramidal Neurons and Stellate Cells [70]. The fourth, fifth and sixth layers are dedicated to the association of spatial and non-spatial features because they are essentially composed of Pyramidal Neurons and Horizontal Chandelier Cells and they can send information directly to the Hippocampus [65]. The Grid Cells appear to create an internal map structure of a spatial environment, where each point in the map is displayed in an hexagonal form when applied to a 2D environment [5]. The display of the points and the firing of these cells is determined by the velocity and the head direction of the individual, where the hexagonal shape is maintained at all times as it can be observed in **Figure 3**. The **Figure 3** has images that enhances the hexagonal shape of the Grid Cells and the results that proved its properties.

The velocity and direction, are represented by the Speed Cells and HD cells [5]. The Speed Cells and HD cells are supported by the Conjunctive Cells that discharge their energy potential when a subject is facing a specific direction of movement. Also, it is possible that the Conjunctive Cells are responsible for the encoding of data combinations between Speed Cells, Place Cells and HD cells, allowing them to act as an information integrator [72]. The Grid Cells firing displacement, have a spacing value close to and they are organized an multi-layered form. Each layer represents a different spatial precision of the environment, allowing the increase of certainty, relative to the position of the agent [73].

4.2 Importance of vision and audition in space characterization

As mentioned in Section 1, all the senses have a relative importance in the gathering of the spatial data. For this chapter we will focus on the Vision and Audition.

The Vision and Audition send their information in two different pathways, the Dorsal stream which can indicate the information related to high-level information like the where and how, and the Ventral stream that has low-level data such as the recognizing and identification of objects [74].

Yet, as mentioned in the previous sections, the visual elements are used by the HD cells as inputs used in the determination of the heading. On the other hand, the audition contributes to the spatial characterization because some of the landmarks and spatial identifications are obtained using the auditory system [75]. Also, it is necessary to understand how the sensory cues are converted from the physical element to the neural language and how such data is interpreted by other brain structures [76].

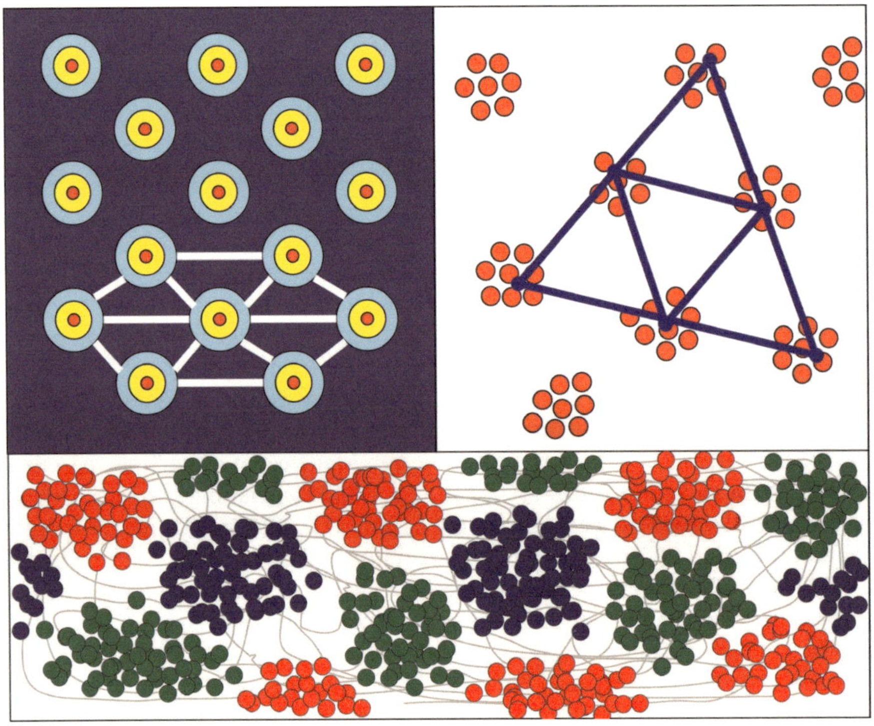

Figure 3.
The Grid Cells main characteristics adapted from [3, 5, 71]. The left image shows the hexagonal shape of the Grid Cells, the right enhances the possible triangle properties. The bottom image is an adaptation of the obtained scientific results from [71].

4.2.1 Visual cortex

The Human Vision System can be seen as two distinct parts. The first part consists in an image capturing system, where the eye and its components are used to gather visual information, and the second part is dedicated to its processing and the interpretation of the acquired data, in the so called Visual cortex.

The Human Eye is mostly formed by: (i) the cornea, which refracts light as it enters the eye, focusing the image on the retina, at the back of the eye [77], (ii) the Aquoeus Hummour that provides nutrition and oxygen to the eye's tissues, as well as maintain the shape of the eye by adjustment of the internals' eye pressure, (iii) the Iris and Pupil which are responsible for the control of the tunnel that guides the light rays [77, 78]. The conversion between the light's electromagnetic waves into neural signals, occurs in the Retina. The Retina has Photo-Receptor cells, Amacrine Cells, Bipolar Cells and Ganglion Cells. The Photo-Receptor Cells are sensitive to light. These are either Rods or Cones, respectively sensitive to the tones of gray and colors at different areas of the retina. There are 3 types of Cones, each one processes and triggers at the presence of the different electromagnetic wavelets like the red, green and blue colors [77].

The information collected from the cones is aggregated with the data from the rods using the Horizontal Cells, which are inhibitory interneurons, that balance the overall activity of the retina and improve the contrast of the visual signal that is sent to the brain. According with [79], it is possible that Horizontal Cells can have some characteristics used in color opponency. Color opponency consists in the creating a new color by suppressing the other wavelets. Such characteristic allows the reduction of redundant information generated by the photo-receptor electrical discharges, avoiding overlap of the cone sensitivities [80].

Also, the Amacrine Cells establish links between the different layers and they could play a role in the set of contrast of the visual image. The data is sent to the Visual Cortex through the Ganglion Cells. They also have a crucial role at the regulation and synchronization of the circadian rhythm [77].

The eye and the Vision Cortex can capture and identify different image aspects, such as the edges of a shadow or a object, the motion of elements, they can distinguish from large to small objects and they have the ability to process large visual fields [81].

The information regarding the recognition of objects, processed in the Visual Cortex, is sent to the Temporal Lobe elements via the Ventral pathway where the Perirhinal Cortex acts as a hub between the Hippocampus and the visual system [82].

From the image capturing by the retina, to the data preparation made by the Visual Cortex, the usage of the Visual data in navigation can be important in the identification of spatial landmarks, the gathering of information present in a cartographic map and they can correct some of the errors made during the path-integration process. Also, the Vision data is used by the Hippocampal Place Cells to encode the specific area in the spatial environment [83].

4.2.2 Auditory cortex

A spatial environment can be characterized by the environmental sound. The Auditory mechanism has similar features with the Visual System. It can be divided into two parts: the first part is dedicated to the capturing of the sound waves and the second part for the processing and interpretation of the information gathered from the sound waves.

The sound waves are captured at the structure called external ear and the sound vibrations are extracted by the tympanic membrane which sends the vibrating rhythms into the cochlea. The cochlea has a set of elements called Hair Cells and they move with a certain vibrating frequency. Such movement causes the discharge of an energy potential, converting the sound's vibration into an neural signal. The electrical impulses are sent to the Auditory Cortex where the sound is classified according with their properties [84].

The Auditory Cortex seem to have several neural populations that react to different sound characteristics. Such statement was supported by several observations of voxels in Functional Magnetic Resonance images [85]. These elements are present in the Primary Auditory Cortex, that identifies and sorts data according to their frequency [86]. The Second Auditory Cortex identifies the sound using the pitch and melody. Also, between the Auditory Cortex and the Speech Cortex is the language processing area, necessary for the extraction of information from the voice [85].

The auditory landmarks can help in the identification and recognition of places, but according with [75] some of these landmarks are quite difficult to gather and sometimes the information acquired needs to be compared with other senses like the Vision. Also, it is stated that the ambiguous information will decrease with the amount of Visual and Auditory data. Since the audio data is stored by its frequency, it is required to locate the position of the sound source from the sound profiles received by the two ears taking into account that it is required to solve the ambiguity and sound input differences between the two ears.

The extraction of spatial features in the sound could be difficult, because of the ears position. The cues obtained using the frontal part and the back part of the head, are equal and such difficult is solved by the moving of the head to other position [87].

The difference of sound arrival can be important to the creation of a auditory spatial map which can lead to the detection of important environmental cues [88], but they need to be fused with other elements, like the Visual data, to separate the environmental data segments [75].

4.3 Encoding of sensory information

This subject lacks consensus among the scientific community. However, there are several hypothesis about how the brain handles the sensory information [89, 90].

One of the theories that has the most consensus consists in fact that the brain encodes the information in a sparse manner [89]. A sparse distributed representation can be described has a representation where not all neurons are active at a given period of time. The sparse representation together with the firing rate of the neuron can be used in the encoding of some of the sensory information like the auditory data. The generalization and pattern completion depend in the sparseness of the data and in the distributed characteristic of the neurons and the less sparse is the system the more information can be encoded and represented. A sparse codification system can be found in different areas of brain like in (i) the Auditory Cortex where the firing rate can be characterized as a binary coding, because the neuron responds to the stimulus in a fire or no fire behavior, (ii) in the Hippocampus and (iii) in the motor sensory cortex [91].

On the other hand, it is not clear how the brain enforces such codification and how this method could lead to generalization when the interactions between neurons are very low. These statements are supported by the fact that the sparse coding is not present in all the areas of the cortex and it is not clear if the brain is a discrete classifier or if it works using continuous regression, making the system totally dependent to the size of the neuronal population [92].

5. Computational methods about hippocampal formation and spatial navigation

Nowadays, due to the increase of technological capabilities it is possible to implement complex structures of neural networks. However, some of the used methods do not have a strong biological justification. Also, nowadays it is possible to simulate some of the current brain's neurological behaviors using certain type of networks. Those methods and strategies include neuronal simulators like Nengo [93] or The Brian Simulator [94] but also with different types of networks like the SNN. There are several models which try to simulate some of the characteristics of the EC, others focus in the Hippocampus and some even use the brain and Hippocampal features applied to navigation.

The following sub-sections start with an analysis with the EC models followed by an analysis of the Hippocampal model where it be given emphasis on their main technological characteristics.

5.1 Entorhinal cortex models

Throughout the years several computer models were developed that tried to replicate some or all elements of the EC. Some of the models focused on certain types of cells like the Grid Cells. The Grid Cells models can be divided by their main characteristics in: (1) Oscillatory-Interference Model, (2) the Attractor Network Model, (3) the Self-Organized Model and (4) the Path-Integration Model [95].

The Oscillatory-Interference method makes use of two different oscillators and their phase difference creates the Grid Cells [95]. The oscillators can be active although it is required for them to have different frequencies. This model can respond to rapid changes in the environment and at the same time they could provide the necessary information for a rapid remapping of Place Cells [96].

The Attractor Network model uses simulated neurons, with different types of connections to represent a position in a space. The firing of the neuron is directly influenced by the behavior of the neighboring neurons. However, to recreate the functions and behaviors of the EC, this network must have a element which detects and recognizes borders/obstacles to limit the initialization of hyper-parameters, avoiding the data overfit problem [95].

A Self-Organized mechanism states that the data has to be separated by a 60-degree angle. These model makes uses of a special type of cells called Stripe Cells [97]. The Stripe Cells send to the Grid Cells, the speed information and they constrict the formation of Grid Cells by maintaining a constant hexagonal pattern displacement [95]. Those cells have different firing behaviors and the intersection of those elements create the Grid Cells. These approach is sensitive to trigonometric properties and it can represent different ranges of spatial scales [98].

The Path-Integration method states that Grid Cell's formation only depend from the velocity and direction of the agent. Also, the displacement of Grid Cells is derived by the multiplication of the previous elements with a timing factor [95].

According to [71], it is possible to recreate the EC Grid Cells using simple mathematical properties. By separating the hexagonal shape into triangles and with a specific type of Gray code, the Grid Cells are formed and characterized using a Triangular Coordinate System. This method has as its main characteristic the ability to define specific degrees of precision, because it behaves in a multi-layered form, like the real Grid Cells. Also, it assembles a property of similarity where the code of the near elements has the lowest possible difference, making this method suitable for navigation and to encode multidimensional data.

5.2 Hippocampus models

When the Hippocampus is referenced, there are several computational models which enhance some of the characteristics mentioned in the Introduction section. The NeuroSLAM [99] and RatSLAM [100] are some of the models based in the characteristics of the Hippocampal Formation.

The RatSLAM method consists in a mapping algorithm based in the characteristics of the rat's Hippocampal Formation system. This method uses the Competitive Attractor Network (CAN) to convert the Visual information to a position in a space. The CAN can have different structures and the idea consists in exciting a neuron with its neighbors, and to inhibit the distant neurons, with the ability to handle small amounts of noisy data [101]. Also, this method uses landmarks for spatial guidance and it is not dependent of a special grid system like the Cartesian method [100]. In addition, the core of the RatSLAM, had been used in the development of other neural based navigational models, like the OpenRatSLAM, that has a modular structure which is easy to implement in robots [102], and the DolphinSLAM where the properties of RatSLAM, are used in underwater environments [103].

The NeuroSLAM is a spatial neural inspired model that can be implemented in 2D and 3D environments. This method uses some of the previously mentioned biological elements like the Place Cells, HD cells and Grid Cells. The Grid Cells maintain their hexagonal displacement. They represent the position and give the metric information used in the path integration. The HD cells provide information regarding the azimuth and the heading direction of the movement. The networks for this model are made using Multidimensional Continuous Attractor Network (MD-CAN) that, with time, achieve a stable state, when the right network parameters are used [101]. Each unit has a continuous activation value between zero and one. With the simulated neurons, this model makes use of the Local View Cells which are able to resets the errors gathered during the path integration process based on the rotation data provided by the 3D visual odometry data [99].

The Neural Engineering Framework (NEF) is a model which can be applied to navigational problems. The NEF allows the simulation of biological neural models applied to cognition and other brain tasks. This can be used model visual attention, reinforcement learning and other cognitive jobs [104]. This framework allows a better understanding of the brain mechanisms using fast computational resources. The neural activity uses an tuning curve as the firing rule, which is described in Eq. (1)

$$a_i = G(a_i e_i{}^* x + b_i) \tag{1}$$

where the G is a neural model, a_i is the gain parameter, b_i is the background bias and the x is the neural activity. There are several available neural models like the rate-based sigmoidal neurons, spiking Leaky-Integrate-and-Fire (LIF) neurons and other models [104]. The NEF properties can be used in programming libraries like the Nengo where the neural models are created with programming languages like Python or Java and it allows the visualization of the model behavior using Graphical User Interface tools. In terms of hardware it is possible to use FPGAs to create physical neural models, and it is also possible to convert from SNN's to Artificial Neural Networks (ANN) [105].

SNNs are a special type of network that try to recreate the neural structure of the brain. They work in a sparse manner and in a continuous time format. There are several SNN neural models like the Hodgkin-Huxley (HH) model when the discharge of sodium and potassium is simulated inside the digital neurons or the Izhikevich model which combines the characteristics of the LIF, creating a more efficient digital neuron [106]. The NEF and SNN models do not have a direct connection to the previously mentioned SLAM methods. However, with the right neural configurations, those methods can solve SLAM problems because those methods are simulations of the brain's neural dynamics [107].

6. Discussion and conclusions

Inside the brain's Hippocampal structure there are some neurological tools that allow navigation. Their discovery led to some questions, namely: (i) what are the functions of the Hippocampal Formation and (ii) what is the correlation between the Hippocampus and Navigation?

The first studies about the Hippocampus have shown that it was essential to the formation of short-term memories, as was observed and reported for a patient that had to remove this neuronal structure to solve a problem related to seizures [1]. After the removal, he repeated routine tasks, as if it was the first time ever. He could watch the same movie repetitively, without remembering any scene, actor, or any detail. A similar thing happened with people and places. So, the Hippocampus has shown to allow the identification of places and persons, meaning that it can give the context to a newly formed event [4]. Later studies and researches have made it clear that specific types of neurons, named Place Cells with the help of the EC Grid Cells [5], allow the identification of locations, based on environmental characteristics [7].

Even with the map, we need to have a sense of our current position, which is given by the EC [72], and to have the ability to find new or previously known areas based on the data patterns, which is the function of the Hippocampus [108]. In the Hippocampus the senses, including Vision and Audition, allow the identification of the place, like the SLAM approach which uses the spatial landmarks as a reference to determine the location and the position of a agent. Like the SLAM methods the Hippocampal Formation must have structures that can work with incomplete data. One of those structures is the CA3 area which is composed by recurrent network, that works as a memory element, which can complete some of the missing data [45]. The CA areas and the other Hippocampal elements, process the sensory data and complete the pattern that leads to the identification the current position in an environment.

The Hippocampal Formation is not the only part of the brain that is capable of processing spatial data. Recent works have discovered that the Retrosplenial Cortex, allows the processing of spatial landmarks, creating a spatial schema of the environment, with the possibility to calculate future directions [109].

The Retrosplenial Cortex has a connection to some elements of the Hippocampal Formation. The 29th and 30th areas have a higher neural density which could indicate that the Retrospleanial Cortex process hippocampal related information [110]. Its malfunction could lead to disorientation, because of the lack of ability to pair and process spatial data [109, 110].

Also, it is necessary to further analyze how other senses can contribute to the navigation tasks. "The olfactory input to the perirhinal cortex and the parahippocampal cortex, can provide information which can change the place cell mapping configuration by creating new spatial landmarks [111]. However, the landmark, starts to lose strength due to the odor habituation mechanisms that occur inside the olfactory bulb [112]. In addition, the somatosensory, receives the information from different types of senses and pathways. It passes or holds the sensory information according with the received stimulus [113].

The set of tools created by mankind to navigate in space, have only recently evolved to use non-biological Grid Cells, as defined in discrete global grid systems, that somehow approximate the latices used in the Entorhinal Cortex. Discrete global grid systems currently support the indexing of assets across the globe, allowing a more adequate partitioning and aggregation of the earth into logical structures that take into account the heterogeneity of the scales of the associated geospatial data.

With the increase of information regarding the neuronal navigation capabilities, there are some methods that bridge neuroscience and SLAM. As previously mentioned the NeuroSLAM [99], and RatSLAM [100] can be considered good approaches to solve SLAM using biological based methods. However, due to the complexity of the CAN networks, they cannot handle large quantities of noisy data. Meaning that the data has to be cleaned, which could increase the time of response. Also the MD-CAN networks reveal some instability when there are some changes in the network parameters. Such changes cause a drift in the attractor that leads to the degradation of the results [101].

Also, the RatSLAM uses some cells that have not a direct correlation with the Hippocampal Formation, like the Pose Cells that represent the location and orientation of an agent [100]. In addition, the NeuroSLAM considers an intermediary agent to facilitate the communication between the system and neuromorphic devices, which is not required when using SNNs [99].

On the other hand, the Triangular Coordinate System [71], can handle certain amounts of noise in the data using simple mathematical operations, without using any special equipment like the neuromorphic devices. Yet, this model does not address a codification for 3D environments and the hexagons seem to a have a perfect hexagonal shape, which contrasts with what is observed in some neurobiological studies.

Non-biological based SLAM approaches that were previously mentioned can already be found in some house appliances like in autonomous vacuum cleaners. So does it make sense to create a biological based artificial Hippocampal Formation structure, in robots?

The answer to this question seems to be difficult, because the SNN has a different structure and there are certain aspects that require attention. The first aspect is in the training method. It is different from the ANN because SNNs have special characteristics in the spike process, meaning that the backpropagation mechanisms cannot be directly applied. The second aspect comes from the fact that the SNNs need a computer simulated program. According with [114], it is necessary to understand how such networks could work with high performance computer elements, like the neuromorphic devices. On the other hand, the SNN would not need a large energy storage units and the robot could make difficult tasks in a more efficient way due to the network's performance capabilities [114].

As mentioned in the SLAM section, one of the steps of the SLAM approach is the navigation planning and according with [115], it seems that the Hippocampus Place Cells and the EC Grid Cells, can be used for conceptual learning, due to their capacity to organize information and to predict future states.

With the analysis of the state of the art matter, it is possible to say that the Hippocampal Formation can have the required elements for the task, because it can create a data representation of the place, based on the available environmental characteristics, allowing the understanding of the current location and which are the next available steps.

During this century, neuroscientists and engineers have been trying to close the gap between these fields with a better knowledge that could clarify some of the existing doubts about the functioning of the brain, and with new technological approaches. With the junction of the neuroscience to the robotic field, it may be possible to develop new technologies which are more adapted to the people's necessities and requirements to solve their daily challenges.

Abbreviations

ANN	Artificial Neural Networks
CA	Cornu Ammontis
CAN	Competitive Attractor Network
DG	Dentate Gyrus
GPS	Global Positioning System
HD	Head Direction
LEC	Lateral Entorhinal Cortex

LIF	Leaky-Integrate-and-Fire
MEC	Medial Entorhinal Cortex
MD-CAN	Multidimensional Continuous Attractor Network
NEF	Neural Engineering Framework
SLAM	Simultaneous Localization and Mapping
SNN	Spiking Neural Network

Author details

André Pedro[1]*, Jânio Monteiro[2] and António João Silva[2]

1 Faculdade de Ciências e Tecnologia, Universidade of Algarve, Faro, Portugal

2 Instituto Superior de Engenharia, Universidade do Algarve, Faro, Portugal

*Address all correspondence to: a52193@ualg.pt

References

[1] William Beecher Scoville and Brenda Milner. Loss of recent memory after bilateral hippocampal lesions. Journal of Neurology, Neurosurgery, and Psychiatry. 1957;**20**(1):11

[2] Zhu Y, Gao H, Tong L, Li ZL, Wang L, Zhang C, et al. Emotion regulation of hippocampus using real-time fmri neurofeedback in healthy human. Frontiers in Human Neuroscience. 2019;**13**:242

[3] Moser M-B, Rowland DC, Moser EI. Place cells, grid cells, and memory. Cold Spring Harbor Perspectives in Biology. 2015;7(2):a021808

[4] Teyler TJ, Rudy JW. The hippocampal indexing theory and episodic memory: Updating the index. Hippocampus. 2007;**17**(12): 1158-1169

[5] Rowland DC, Roudi Y, Moser M-B, Moser EI, et al. Ten years of grid cells. Annual Review of Neuroscience. 2016; **39**:19-40

[6] Bush D, Barry C, Manson D, Burgess N. Using grid cells for navigation. Neuron. 2015;**87**(3):507-520

[7] Go MA, Rogers J, Gava GP, Davey CE, Seigfred Prado Y, Liu, and Simon R Schultz. Place cells in head-fixed mice navigating a floating real-world environment. Frontiers in Cellular Neuroscience. 2021;**15**:618658

[8] Newcombe NS. Navigation and the developing brain. Journal of Experimental Biology. 2019;**222** (Suppl_1):jeb186460

[9] Getting IA. Perspective/navigation-the global positioning system. IEEE Spectrum. 1993;**30**(12):36-38

[10] McCarthy DD, Seidelmann PK. Time: From Earth Rotation to Atomic Physics. Cambridge, United Kingdom: Cambridge University Press; 2018

[11] Durrant-Whyte H, Bailey T. Simultaneous localization and mapping: Part i. IEEE Robotics & Automation Magazine. 2006;**13**(2):99-110

[12] Sadler DH. Lunar distances and the nautical almanac. Vistas in Astronomy. 1976;**20**:113-121

[13] Roston GP, Krotkov EP. Dead Reckoning Navigation for Walking Robots. Technical Report. Pittsburgh, United States: Carnegie-Mellon University Pittsburgh PA Robotics Institute; 1991

[14] Richey M. The navigational background to 1492. The Journal of Navigation. 1992;**45**(2):266-284

[15] Wagner J, Sorg HW. The bohnenberger machine. Gyroscopy and Navigation. 2010;**1**(1):73-78

[16] MJP Vis. History of the Mercator projection [B.S. thesis]. Heidelberglaan, Utrecht, Netherlands: Utrecht University; 2018

[17] Akram M, Khiyal H, Ahmad M, Abbas S. Decision Tree for Selection Appropriate Location Estimation Technique of GSM Cellular Network. In: International conference on engineering & emerging technology, Lahore, Pakistan. March 2014

[18] Šavrič BJ, Patterson T, Petrovič D, Hurni L. A polynomial equation for the natural earth projection. Cartography and Geographic Information Science. 2011;**38**(4):363-372

[19] Peterson PR. Discrete global grid systems. In: International Encyclopedia of Geography: People, the Earth, Environment and Technology: People, the Earth, Environment and Technology. 2016. pp. 1-10

[20] Bush I, Riscaware L. Openeaggr Software Design Document. GitHub-Riskaware Ltd. 2017. Available from: https://github.com/riskaware-ltd/open-eaggr [Accessed: 22 June 2019]

[21] Goodchild F Michael, Kimerling A Jon. Discrete Global Grids: A Web Book. 2002.

[22] Mahony R, Hamel T, Trumpf J. An homogeneous space geometry for simultaneous localisation and mapping. Annual Reviews in Control. 2021;**51**: 254-267

[23] Huang B, Zhao J, Liu J. A survey of simultaneous localization and mapping. arXiv preprint arXiv:1909.05214, 2019

[24] Zhou Y, Tuzel O. Voxelnet: End-to-end learning for point cloud based 3d object detection. In: Proceedings of the IEEE Conference on Computer Vision and Pattern Recognition. Los Alamitos, CA, USA: IEEE Computer Society. 2018. pp. 4490-4499

[25] Khairuddin A, R, Talib MS, Haron H. Review on simultaneous localization and mapping (slam). In: 2015 IEEE International Conference on Control System, Computing and Engineering (ICCSCE). Penang, Malaysia: IEEE; 2015. pp. 85-90

[26] Stasse O, Davison AJ, Sellaouti R, Yokoi K. Real-time 3d slam for humanoid robot considering pattern generator information. In: 2006 IEEE/RSJ International Conference on Intelligent Robots and Systems. Beijing, China: IEEE; 2006. pp. 348-355

[27] Hening S, Ippolito CA, Krishnakumar KS, Stepanyan V, Teodorescu M. 3d lidar slam integration with gps/ins for uavs in urban gps-degraded environments. In: AIAA Information Systems-AIAA Infotech@ Aerospace. 2017. p. 0448

[28] Chiang K-W, Tsai G-J, Li Y-H, Li Y, El-Sheimy N. Navigation engine design for automated driving using ins/gnss/3d lidar-slam and integrity assessment. Remote Sensing. 2020; **12**(10):1564

[29] Grisetti G, Stachniss C, Burgard W. Improved techniques for grid mapping with rao-blackwellized particle filters. IEEE Transactions on Robotics. 2007; **23**(1):34-46

[30] Carlone L, Aragues R, Castellanos JA, Bona B. A linear approximation for graph-based simultaneous localization and mapping. In: Robotics: Science and Systems. Vol. 7. Cambridge, Massachusetts, USA: The MIT Press; 2012. pp. 41-48

[31] Zhang J, Singh S. Loam: Lidar odometry and mapping in real-time. In: Robotics: Science and Systems. Vol. 2. Berkeley, CA, Cambridge, Massachusetts, USA: MIT Press; 2014. pp. 1-9

[32] Deschaud J-E. Imls-slam: Scan-to-model matching based on 3d data. In: 2018 IEEE International Conference on Robotics and Automation (ICRA). Brisbane, QLD, Australia: IEEE; 2018. pp. 2480-2485

[33] Bichen W, Wan A, Yue X, Keutzer K. Squeezeseg: Convolutional neural nets with recurrent crf for real-time road-object segmentation from 3d lidar point cloud. In: 2018 IEEE International Conference on Robotics

and Automation (ICRA). Brisbane, QLD, Australia: IEEE; 2018. pp. 1887-1893

[34] Mur-Artal R, Tardós JD. Orb-slam2: An open-source slam system for monocular, stereo, and rgb-d cameras. IEEE Transactions on Robotics. 2017; **33**(5):1255-1262

[35] Rublee E, Rabaud V, Konolige K, Bradski G. ORB: An efficient alternative to SIFT or SURF. In: 2011 International Conference on Computer Vision. Barcelona, Spain: IEEE; 2011. pp. 2564-2571

[36] Sumikura S, Shibuya M, Sakurada K. Openvslam: A versatile visual slam framework. In: Proceedings of the 27th ACM International Conference on Multimedia. 2019. pp. 2292-2295

[37] Qin T, Li P, Shen S. Vins-mono: A robust and versatile monocular visual-inertial state estimator. IEEE Transactions on Robotics. 2018;**34**(4): 1004-1020

[38] Qin T, Shen S. Online temporal calibration for monocular visual-inertial systems. In: 2018 IEEE/RSJ International Conference on Intelligent Robots and Systems (IROS). Madrid, Spain: IEEE; 2018. pp. 3662-3669

[39] DeTone D, Malisiewicz T, Rabinovich A. Toward geometric deep slam. arXiv preprint arXiv:1707.07410. 2017

[40] Dai A, Ritchie D, Bokeloh M, Reed S, Sturm J, Nießner M. Scancomplete: Large-scale scene completion and semantic segmentation for 3d scans. In: Proceedings of the IEEE Conference on Computer Vision and Pattern Recognition. Salt Lake City, Utah, United States: IEEE; 2018. pp. 4578-4587

[41] Xiang Y, Fox D. Da-rnn: Semantic mapping with data associated recurrent neural networks. arXiv preprint arXiv: 1703.03098. 2017

[42] Silveira L, Guth F, Fisher D, Codevilla F, Drews P, Botelho S. Biological inspired system for localization and mapping in underwater environments. In: 2013 OCEANS-San Diego. San Diego, CA, USA: IEEE; 2013. pp. 1-6

[43] Barry C, Lever C, Hayman R, Hartley T, Burton S, O'Keefe J, et al. The boundary vector cell model of place cell firing and spatial memory. Reviews in the Neurosciences. 2006;**17**(1–2):71-98

[44] Weilbächer RA, Gluth S. The interplay of hippocampus and ventromedial prefrontal cortex in memory-based decision making. Brain Sciences. 2016;**7**(1):4

[45] Cherubini E, Miles R. The ca3 Region of the Hippocampus: How Is it? What Is it for? How Does it Do it?, Lausanne, Switzerland, 2015;**9**. DOI: 10.3389/fncel.2015.00019. Available from: http://journal.frontiersin.org/Article/10.3389/fncel.2015.00019/abstract

[46] Umschweif G, Greengard P, Sagi Y. The dentate gyrus in depression. European Journal of Neuroscience. 2021; **53**(1):39-64

[47] Urban NN, Henze DA, Barrionuevo G. Revisiting the role of the hippocampal mossy fiber synapse. Hippocampus. 2001;**11**(4):408-417

[48] Senzai Y. Function of local circuits in the hippocampal dentate gyrus-ca3 system. Neuroscience Research. 2019; **140**:43-52

[49] Molitor RJ, Sherrill KR, Morton NW, Miller AA, Preston AR. Memory

reactivation during learning simultaneously promotes dentate gyrus/ ca2, 3 pattern differentiation and ca1 memory integration. Journal of Neuroscience. 2021;**41**(4):726-738

[50] Kesner RP. A process analysis of the ca3 subregion of the hippocampus. Frontiers in Cellular Neuroscience. 2013; 7:78

[51] Pata DS, Escuredo A, Lallée S, Verschure PFMJ. Hippocampal based model reveals the distinct roles of dentate gyrus and ca3 during robotic spatial navigation. In: Conference on Biomimetic and Biohybrid Systems. Milan, Italy: Springer; 2014. pp. 273-283

[52] Hitti FL, Siegelbaum SA. The hippocampal ca2 region is essential for social memory. Nature. 2014;**508**(7494): 88-92

[53] Dudek SM, Alexander GM, Farris S. Rediscovering area ca2: Unique properties and functions. Nature Reviews Neuroscience. 2016;**17**(2):89-102

[54] MacDonald CJ, Tonegawa S. Crucial role for ca2 inputs in the sequential organization of ca1 time cells supporting memory. Proceedings of the National Academy of Sciences. 2021;**118**(3): e2020698118

[55] Deshmukh SS. Distal ca1 maintains a more coherent spatial representation than proximal ca1 when local and global cues conflict. Journal of Neuroscience. 2021;**41**(47):9767-9781

[56] Voneida TJ, Vardaris RM, Fish SE, Reiheld CT. The origin of the hippocampal commissure in the rat. The Anatomical Record. 1981;**201**(1): 91-103

[57] Harland B, Contreras M, Souder M, Fellous J-M. Dorsal ca1 hippocampal place cells form a multi-scale representation of megaspace. Current Biology. 2021;**31**(10):2178-2190

[58] Geiller T, Fattahi M, Choi J-S, Royer S. Place cells are more strongly tied to landmarks in deep than in superficial ca1. Nature Communications. 2017;**8**(1):14531

[59] Poulter S, Hartley T, Lever C. The neurobiology of mammalian navigation. Current Biology. 2018;**28**(17):R1023-R1042

[60] Lever C, Burton S, Jeewajee A, O'Keefe J, Burgess N. Boundary vector cells in the subiculum of the hippocampal formation. Journal of Neuroscience. 2009;**29**(31):9771-9777

[61] Deshmukh SS, Knierim JJ. Influence of local objects on hippocampal representations: Landmark vectors and memory. Hippocampus. 2013;**23**(4): 253-267

[62] Kitanishi T, Umaba R, Mizuseki K. Robust information routing by dorsal subiculum neurons. Science Advances. 2021;**7**(11):eabf1913

[63] Kazanovich YB, Evgen'evich Mysin I. How animals find their way in space. Experiments and modeling. Mathematical Biology and Bioinformatics. 2018;**13**(Suppl):132-161

[64] Kim M, Maguire EA. Encoding of 3d head direction information in the human brain. Hippocampus. 2019;**29**(7): 619-629

[65] Canto C, B, Wouterlood FG, Witter MP. What does the anatomical organization of the entorhinal cortex tell us? Neural Plasticity. 2008;**2008**

[66] Tsao A, Sugar J, Li L, Wang C, Knierim JJ, Moser M-B, et al. Integrating

time from experience in the lateral entorhinal cortex. Nature. 2018; **561**(7721):57-62

[67] Rolls ET, Mills P. The generation of time in the hippocampal memory system. Cell Reports. 2019;**28**(7):1649-1658

[68] Woodruff AR, Anderson SA, Yuste R. The enigmatic function of chandelier cells. Frontiers in Neuroscience. 2010;**4**:201

[69] Rowland DC, Obenhaus HA, Skytøen ER, Zhang Q, Kentros CG, Moser EI, et al. Functional properties of stellate cells in medial entorhinal cortex layer ii. eLife. 2018;**7**:e36664

[70] Jones RSG. Entorhinal-hippocampal connections: A speculative view of their function. Trends in Neurosciences. 1993; **16**(2):58-64

[71] Monteiro J, Pedro A, Silva AJ. A gray code model for the encoding of grid cells in the entorhinal cortex. Neural Computing and Applications. 2022; **34**(3):2287-2306

[72] Tukker JJ, Beed P, Brecht M, Kempter R, Moser EI, Schmitz D. Microcircuits for spatial coding in the medial entorhinal cortex. Physiological Reviews. 2022;**102**(2):653-688

[73] Gay S, Le Run K, Pissaloux E, Romeo K, Lecomte C. Towards a predictive bio-inspired navigation model. Information. 2021;**12**(3):100

[74] Rauschecker JP. Where, when, and how: Are they all sensorimotor? Towards a unified view of the dorsal pathway in vision and audition. Cortex. 2018;**98**: 262-268

[75] Jetzschke S, Ernst MO, Froehlich J, Boeddeker N. Finding home: Landmark ambiguity in human navigation. Frontiers in Behavioral Neuroscience. 2017;**11**:132

[76] Zavitz E, Price NSC. Understanding sensory information processing through simultaneous multi-area population recordings. Frontiers in Neural Circuits. 2019;**12**:115

[77] Zhu J, Zhang E, Del Rio-Tsonis K. Eye Anatomy. eLS. Hoboken, New Jersey, USA: Wiley Online Library; 2012

[78] Bouma H. Size of the static pupil as a function of wave-length and luminosity of the light incident on the human eye. Nature. 1962;**193**(4816): 690-691

[79] Twig G, Levy H, Perlman I. Color opponency in horizontal cells of the vertebrate retina. Progress in Retinal and Eye Research. 2003;**22**(1):31-68

[80] Lee T-W, Wachtler T, Sejnowski TJ. Color opponency is an efficient representation of spectral properties in natural scenes. Vision Research. 2002; **42**(17):2095-2103

[81] Sousa NPPA. Neural encoding models in natural vision. Porto, Portugal: Faculdade de Engenharia da Universidade do Porto; 2013

[82] Turk-Browne NB. The hippocampus as a visual area organized by space and time: A spatiotemporal similarity hypothesis. Vision Research. 2019;**165**: 123-130

[83] Ekstrom AD. Why vision is important to how we navigate. Hippocampus. 2015;**25**(6):731-735

[84] Musiek FE, Baran JA. The Auditory System: Anatomy, Physiology, and Clinical Correlates. San Diego, CA: Plural Publishing; 2018

[85] Norman-Haignere SV, Feather J, Boebinger D, Brunner P, Ritaccio A, McDermott JH, et al. A neural population selective for song in human auditory cortex. Current Biology. 2022; **32**(7):1470-1484

[86] Weinberger NM. Auditory associative memory and representational plasticity in the primary auditory cortex. Hearing Research. 2007;**229**(1–2):54-68

[87] McAnally KI, Martin RL. Sound localization with head movement: Implications for 3-d audio displays. Frontiers in Neuroscience. 2014;**8**:210

[88] King AJ, Schnupp JWH, Doubell TP. The shape of ears to come: Dynamic coding of auditory space. Trends in Cognitive Sciences. 2001;**5**(6):261-270

[89] Rolls ET, Treves A. The neuronal encoding of information in the brain. Progress in Neurobiology. 2011;**95**(3): 448-490

[90] Godenzini L, Alwis D, Robertas Guzulaitis S, Honnuraiah GJS, Palmer LM. Auditory input enhances somatosensory encoding and tactile goal-directed behavior. Nature Communications. 2021;**12**(1):1-14

[91] Olshausen BA, Field DJ. Sparse coding of sensory inputs. Current Opinion in Neurobiology. 2004;**14**(4): 481-487

[92] Spanne A, Jörntell H. Questioning the role of sparse coding in the brain. Trends in Neurosciences. 2015;**38**(7): 417-427

[93] Bekolay T, Bergstra J, Hunsberger E, DeWolf T, Stewart TC, Rasmussen D, et al. Nengo: A python tool for building large-scale functional brain models. Frontiers in Neuroinformatics. 2014;**7**:48

[94] Goodman DFM, Brette R. The brain simulator. Frontiers in Neuroscience. Lausanne, Switzerland: Frontiers Media SA; 2009:26

[95] Giocomo LM, Moser M-B, Moser EI. Computational models of grid cells. Neuron. 2011;**71**(4):589-603

[96] Burgess N. Grid cells and theta as oscillatory interference: Theory and predictions. Hippocampus. 2008;**18**(12): 1157-1174

[97] Pilly PK, Grossberg S. Spiking neurons in a hierarchical self-organizing map model can learn to develop spatial and temporal properties of entorhinal grid cells and hippocampal place cells. PLoS One. 2013;**8**(4):e60599

[98] Mhatre H, Gorchetchnikov A, Grossberg S. Grid cell hexagonal patterns formed by fast self-organized learning within entorhinal cortex. Hippocampus. 2012;**22**(2):320-334

[99] Fangwen Y, Shang J, Youjian H, Milford M. Neuroslam: A brain-inspired slam system for 3d environments. Biological Cybernetics. 2019;**113**(5): 515-545

[100] Milford MJ, Wyeth GF, Prasser D. Ratslam: A hippocampal model for simultaneous localization and mapping. In: IEEE International Conference on Robotics and Automation, 2004. Proceedings. ICRA'04. 2004. Vol. 1. IEEE; 2004. pp. 403-408

[101] Latham PE, Deneve S, Pouget A. Optimal computation with attractor networks. Journal of Physiology-Paris. 2003;**97**(4–6):683-694

[102] Ball D, Heath S, Wiles J, Wyeth G, Corke P, Milford M. Openratslam: An open source brain-based slam system. Autonomous Robots. 2013;**34**:149-176

[103] Zaffari GB, dos Santos MM, Duarte AC, Fernandes DDA, Silvia SDCB. Exploring the dolphinslam's parameters. In: OCEANS 2016-Shanghai. Shanghai, China: IEEE; 2016. pp. 1-5

[104] Stewart TC. A Technical Overview of the Neural Engineering Framework. University of Waterloo. London, United Kingdom: AISB Quartely. Vol. 110. 2012

[105] DeWolf T, Jaworski P, Eliasmith C. Nengo and low-power ai hardware for robust, embedded neurorobotics. Frontiers in Neurorobotics. 2020;**14**: 568359

[106] Yamazaki K, Vo-Ho V-K, Bulsara D, Le N. Spiking neural networks and their applications: A review. Brain Sciences. 2022;**12**(7):863

[107] Galluppi F, Conradt J, Stewart T, Eliasmith C, Horiuchi T, Tapson J, et al. Spiking ratslam: Modeling rat hippocampus place, grid and boarder cells in spiking neural hardware.

[108] Maurer AP, Nadel L. The continuity of context: A role for the hippocampus. Trends in Cognitive Sciences. 2021; **25**(3):187-199

[109] Mitchell AS, Czajkowski R, Zhang N, Jeffery K, Nelson AJD. Retrosplenial cortex and its role in spatial cognition. Brain and Neuroscience Advances. 2018;**2**:2398212818757098

[110] Vann SD, Aggleton JP, Maguire EA. What does the retrosplenial cortex do? Nature Reviews Neuroscience. London, UK: Nature Publishing Group; 2009; **10**(11):792-802

[111] Fischler-Ruiz W, Clark DG, Joshi NR, Devi-Chou V, Kitch L, Schnitzer M, et al. Olfactory landmarks and path integration converge to form a cognitive spatial map. Neuron. 2021; **109**(24):4036-4049

[112] Chaudhury D, Manella L, Arellanos A, Escanilla O, Cleland TA, Linster C. Olfactory bulb habituation to odor stimuli. Behavioral Neuroscience. 2010;**124**(4):490

[113] ten Donkelaar HJ, ten Donkelaar HJ, Broman J, van Domburg P. The somatosensory system. In: Clinical Neuroanatomy: Brain Circuitry and its Disorders. 2020. pp. 171-255

[114] Bing Z, Meschede C, Röhrbein F, Huang K, Knoll AC. A survey of robotics control based on learning-inspired spiking neural networks. Frontiers in Neurorobotics. 2018;**12**:35

[115] Mok RM, Love BC. A non-spatial account of place and grid cells based on clustering models of concept learning. Nature Communications. 2019;**10**(1): 5685

Section 3

Hippocampal Role in Skill Acquisition

Chapter 4

The Fundamental Role of Memory Systems in Children's Writing Skills

Cecilia Beatriz Moreno and Ángel Javier Tabullo

Abstract

Academic skill learning involves different memory systems. Procedural memory needs repetition, while episodic memories are formed from single events and concepts are stored as associative networks within semantic memory. During writing, various cognitive, phonological and motor processes are executed through working memory; whereas long-term memory provides the knowledge that will be recovered during textual production. Proper functioning of these memory systems -and neural substrates such as hippocampus and temporal cortical areas- are related to effectiveness of composing a text. Recovery of stored knowledge is involved in the course of expressive fluency, allowing the integration of the semantic components. Children who can divide attention and control processes through working memory, are more effective in writing text. During writing, working memory manipulates and keeps linguistic symbols online; the phonological loop admits and retains verbal information and performs a review that allows preserving the representations by commanding the lexical, syntactic and semantic processes. In this chapter, we will refer to the theoretical contribution of long-term and working memory systems to children's writing skills, we will examine the neural substrates and cognitive development of these systems and we will present empirical evidence of their role in high and low-level components of the writing process.

Keywords: writing ability, writing skills, narrative texts, working memory, long term memory, school children

1. Introduction

For some time now, neurosciences have provided evidence of the link between memory systems' development and school learning. Schooling requires children to learn a wide range of academic contents as skills (from reading, writing and math to natural and social sciences), and each of them involves different memory systems with specific developmental trajectories and organization principles. For instance, procedural memories are formed through practice and repetition, while a significant event needs only one occurrence to be stored in episodic memory. On the other hand, concepts are learned as associations maps and stored within semantic memory [1].

Learning to read requires training the visual system to recognize graphemic patterns and connecting these orthographic inputs with phonological, semantic and syntax representations distributed through the brain's language networks. Reading a text engages coordinated processes of word decoding and language comprehension to transform incoming visual input into a series of increasingly complex mental representations: from word and sentence meanings to macrostructure and global aspects of discourse [2]. During reading, semantic contents and linguistic knowledge are evoked by visual input and retrieved from long-term memory, while working memory provides a workplace for integrating this information and building mental models [3]. In turn, writing can be described as the combination of two processes: text generation (or ideation) and transcription (essentially, spelling and handwriting [4, 5]. Writing also requires searching and retrieving information from long-term memory systems, such as grammar, text genre, world knowledge and the representation of rhetorical problems, among others [6]. It has been proposed that working memory plays a central role in planning, composing and reviewing the text [4]. Working memory receives an information flow from many cognitive and linguistic processes, such as the display of phonological resources. During text production, efficient writers achieve an adequate integration of the task context and the resulting text by combining previously stored knowledge with writing planning and the textualization act itself [6].

Significant associations have been found between children memory systems' performance and their text composition skills. Using the available information, children who displayed more developed cognitive processes, dividing attention and self-regulation through working memory, were more effective in writing a narrative text [7]. In the following section, we will refer to the neuropsychological development of these memory systems during the school years before turning to empirical evidence of their contribution to writing skills.

2. Neuropsychological development of memory systems

According to neuropsychological models of cognitive development, the frontal area undergoes a long period of combined synaptogenesis and synaptic pruning, which reaches its maximum activity between 6 and 12 years of age [5, 7]. It is well known that the functions of the frontal lobes include attentional sustaining, planning, organizing, and the use of strategies. Frontal gray matter increases during childhood, peaking in early adolescence, and then gradually declines. Meanwhile, temporal lobes are engaged in language, emotion, and memory processes, and temporal gray matter peaks in late adolescence. Additionally, the hippocampus, located in the medial temporal lobe (MTL), is critical for long-term memory storage and retrieval. There is also evidence that the capacity of these memory systems changes vastly between the ages of 4 and 18 years. However, there's still an ongoing debate regarding the relations between the observed morphometric brain changes and the developmental trajectories of these cognitive abilities [8, 9].

De Haan et al. postulated that early damage to hippocampus impedes the normal neural development for memory. They propose a developmental sequence in which semantic memory is established first, and then episodic memory system gradually emerges as a function of hippocampal development [10]. Shing et al. [11] suggest that episodic memory is related to the development and interaction of the hippocampus and MTL structures, as their individual functions and connectivity contribute significantly to memory development along childhood and adolescence. Although there is some memory capacity in the newborn, it expands in the first year of life. And for this, the maturation of the MTL circuit and its connection with the frontal lobe is necessary. Memory development may involve regions of the temporal cortex coordinating with the hippocampus in a process of increasing specialization [12]. Neuroimaging studies have shown that the patterns of prefrontal and hippocampal activity during long-term memory tasks vary considerably between children, adolescents and adults, suggesting a differential contribution of these structures to encoding, consolidation and retrieval throughout development [13]. During the encoding stage, information from secondary association cortexes is selected and organized for long-term storage, a process that is guided and supervised by prefrontal cortex. As prefrontal involvement gradually increases from childhood to adolescence, so does the efficacy of this process. Memories are established within long-term memory during the consolidation stage, as a result of hippocampal-cortical interactions that strengthen the memory traces across the associative cortex. Since the hippocampal dentate-gryus and its connectivity are not fully developed, memories are more vulnerable to forgetting in younger children, and considerable differences in autobiographical memory performance can be observed between 4 and 8-year-olds. Prefrontal cortex is also implicated in long-term memory retrieval, where it elaborates and monitors a search strategy to locate the relevant contents within declarative memory. It has been proposed that age-related variations in memory retrieval can be explained in terms of post-natal prefrontal development trajectories, and neuroimaging studies indicate that retrieval-related activity (as well as task performance) increases in prefrontal and parietal regions when comparing children, adolescents and young adults. On the other hand, other studies suggest that hippocampal activity patterns during successful retrieval also differ between children and adults (the latter exhibiting larger activity in the head of the hippocampus, the former in the tail).

Regarding working memory, there is evidence that its components, verbal working memory and visuo-spatial sketchpad show different developmental trajectories. In addition, it seems that visuo-spatial working memory develops at different rates, suggesting that visual and spatial information processing may be supported by different brain subsystems. A variety of factors, such as brain growth, increases in knowledge, strategy use, and speed of information processing, as well as changes in the rate of deterioration of memory traces, would contribute to the differences in this development process [14]. Neuroimaging studies signal the frontal and parietal areas as the neural substrates for visuospatial working memory in children, adolescents, and adults [15]; while verbal memory depends on the activation of the frontal lobe in children and adolescents [16]. Converging neuroimaging evidence indicates that prefrontal and parietal volume and structural connectivity predict the developmental trajectory of working memory performance [17, 18]. Furthermore, frontoparietal activation during working memory tasks has been associated with age-related performance increases in children and adolescents [19, 20]. Some longitudinal studies have revealed that the speed of execution of cognitive processes –such as visuo-perceptual and auditory skills - increases substantially in childhood and then declines in adolescence; and agree that there are various sets of resources that influence the performance of working memory [21, 22].

3. The role of long-term and working memory in children's writing skills: evidence from neuropsychological studies

Along with reading, writing is a distinctively human and highly complex skill of paramount importance for academic achievement (and daily life activities as well) [23, 24].

Writing encompasses both low-level lexical, orthographic and graphomotor processes involved in individual word transcription and high-level processes, such as planning, translating and reviewing [23, 25]. Text redaction requires considering the goal of the message and the intended readers' capacity (their reading skills and previous knowledge on the subject), in order to select and organize the most appropriate syntactic structures and lexical items to convey it. In this way, writing planning engages long-term memory access to retrieve information about world knowledge, topic, text genre, as well as vocabulary, grammar and stylistic rules [7, 23, 25]. In addition, working memory provides a workspace where this flow of information can be integrated and structured to generate the text according to the writer's goals. Not only is working memory involved in manipulating and keeping track of the several representations required to build the text, but it is also necessary to temporarily store syntactic, semantic, lexical and orthographic information while writing its sentences [26]. Working memory also interacts with other executive functions [27] through the writing process, such as cognitive flexibility (to alternate between different goals, strategies and text representations) and inhibitory control (to suppress retrieval of irrelevant information or interference from distracting stimuli) [28, 29].

Since writing skills are gradually learned and improved by practice, a further distinction should be made between novice and more experienced writers. Since novice writers' transcription is not fully developed, their working memory is easily overloaded by orthographic coding or graphomotor processing, thus reducing cognitive resources available for higher-level writing processes. Having automatized transcription, more experienced writers can devote additional cognitive resources to goal-setting, interlocutor awareness, stylistic adequacy and rhetorical concerns [30]. As a result, novice writers tend to adopt a knowledge-telling approach, recalling and transcribing contents from their semantic memory, while expert writers display knowledge-transforming strategies, further elaborating on their retrieved contents to meet their rhetorical goals [31].

While most neuropsychological language models have focused on the contributions of working memory, executive control and semantic memory, a specific role for the hippocampal-dependent declarative memory system in supporting online language processing has been proposed [32]. Duff & Schimdt claim that the hippocampal system is engaged during language comprehension and production, due to its capacity for relational binding, representational integration, flexibility, and maintenance. These capabilities would not only be involved in declarative memory consolidation, but they would also provide support for the integration of multimodal information that takes place while understanding or generating speech. While their hypothesis is mostly grounded on evidence from online language processing deficits in patients with hippocampal amnesia, further support can be found in neuroimaging studies of reading and writing. During a creative writing task, Shah et al., [33], found that motor and visual areas for handwriting were activated, along with cognitive and linguistic areas. A right-lateralized activation pattern was observed in the hippocampus, temporal poles and bilateral posterior cingulate cortex, which was associated with episodic memory retrieval, free-associative and spontaneous cognition, and semantic integration. This is congruent with another recent study, which concluded that hippocampal activity contributed to binding and consolidation of incoming information with global context and world knowledge in expository text comprehension [34]. Furthermore, a series of EEG studies of handwriting and drawing in children and adults found desynchronization effects within the alpha and theta bands that were interpreted as evidence of hippocampal involvement [35, 36].

Considering the theoretical links between working and long-term memory systems and writing skills, it becomes apparent that these processes need to be taken into account to explain their acquisition through development. Nevertheless, despite several lines of research addressing lower-level processes, writing has been considerably understudied when compared to reading, particularly among school-aged children [23, 37]. The need for additional research in the Latin American context is emphasized by the fact that children redaction skills have been declining in the region through the last decade, with around 50% of primary school students showing difficulties to generating meaningful texts [38] and 30% exhibiting low performance when applying stylistic resources and conventions [39]. Therefore, in order to bridge the gap in latin american studies of writing skills, a series of studies have been conducted to examine the contribution of long-term and working memory to high and low-level writing processes among Argentinean primary school children [7, 24].

3.1 The role of long-term memory and working memory in lower-level writing processes (translating)

As we mentioned before, [25] seminal model describes writing in terms of three interactive processes: *planning, translating and reviewing*. Planning implies goal-setting, building and organizing a representation of the text schema, according to these goals. Translating refers to the process of transforming those ideas into written language. While it encompasses transcription (spelling and handwriting), it goes beyond it, since it also requires syntactic and lexical-semantic selection to generate the text's sentences. Berninger et al. [40], distinguished two components in the translating process: text generation and transcription. The former is the transformation of ideas into language representations in the working memory. The latter transforms those representations into written language, through the low-level skills of spelling and handwriting. Finally, *reviewing* requires evaluating the written text to identify errors and eventually revising it to better fulfill the writing goals.

Transcription has been the most widely studied aspect of translating [23, 41, 42]. It is acquired during the first school years, and it remains a reliable predictor of the quality of children's texts through primary school [41, 43]. Transcription is gradually automatized throughout children's development, freeing cognitive resources for higher level processes such as planning and reviewing, and thus enabling more efficient and elaborate writing strategies [44]. Primary school is a crucial period for learning to write. During the middle primary school years, translating processes begin to automatize, while planning and reviewing develop gradually. These writing skills gains seem to be linked to the development of several cognitive abilities, such as long-term memory retrieval, working memory and executive functions and visuospatial skills. While long-term memory provides the semantic content for the text, as well as relevant linguistic and world knowledge, it also stores lexical and syntactic information required for building and organizing sentences [23]. Working memory provides a workspace for planning, translating and reviewing processes, and executive functions allow for self-regulation and strategic management of cognitive resources during writing. It has been shown that working memory, inhibitory control and cognitive flexibility are consistent predictors of writing outcomes among primary school children [44–46]. Visuospatial skills are crucial for transcription graphomotor processes [47], but they can also assist reviewing in more experienced writers [48]. The aim of Moreno et al. 's 2022 [24] study was to examine translating processes (transcription and expressive sentence writing), and their association with long term memory, executive functions and visuospatial skills in 3rd to 5th grade Argentinean children.

The sample study consisted of 168 healthy 8–11-year-old children from the 3rd ($n = 56$), 4th ($n = 59$) and 5th ($n = 53$) primary school year. The children completed the following battery of psycholinguistic and neuropsychological tests:

- Translating writing skills: Transcription and sentence expressive writing were assessed with the Writing Fluency and Writing Samples subtests of the Woodcock-Johnson III Test of Achievement, respectively [46]. The first test requires writing sentences or paragraphs prompted by different stimuli, where the second one requires fast sentence writing in response to pictures. A composite expressive writing score can also be calculated.

- Long-term memory access and retrieval: Long-term memory was assessed with the Retrieval Fluency subtest from the Woodcock-Johnson III battery [46], a semantic verbal fluency task.

- Working memory: the Auditory Working Memory subtest from the Woodcock-Johnson III battery was applied [47].

- Visual attention: Visual attention was evaluated with the Magellan Visual Attention Scale [49, 50].

- Planning: was evaluated with the Porteus Maze Test [51].

- Visuo-spatial skills: Visuospatial organization was examined with the Rey Complex Figure Test [52].

Statistical analysis of writing outcomes indicated that children's writing fluency (but not sentence writing) improved as grade increased ($F = 12.86$, $p < .001$). These developmental effects were also observed for all cognitive functions, since each grade group exhibited better scores than the lower ones (F's > 3.46, p's < .035). Regarding the correlations between translating and cognitive processes, writing fluency was significantly associated with long-term and working memory, planning ($.267 < r < .325$, p's < .001) and visuospatial skills ($rho = .235$, $p < .01$), while sentence writing was specifically associated with working and long-term memory (r's > .160, p's < .05). A composite expressive writing score was calculated from both subtests to examine the contribution of memory and executive processes. A hierarchical regression model explained 18.2% of this expressive writing score (see **Table 1**), indicating working memory, long-term memory and visuospatial skill as predictors ($.146 < \beta < .220$, p's < .05).

The observed effect of grade on transcription processes is consistent with the automatization of orthographic coding that is proposed to take place across elementary education [29, 53], and with the previously reported improvement of writing skills that starts in the fourth grade [44]. Performance also increased with grade for memory systems, visuospatial skill and executive functions (as a consequence of both schooling and neurocognitive development), and the observed correlations point out the contribution of different cognitive systems to writing skills along the primary school years [29, 41, 45].

Models	F	R^2	DR^2	B	Standar Error	b	1-b	f^2
Setp 1	7.965** (1166)	.046	.040				.311	.048
Planning				−.599	.035	.214**		
Spet 2	9.210** (2165)	.100	.090				.739	.111
Planning				.084	.034	.183*		
WM				.036	.011	.236**		
Step 3	8.615** (3164)	.136	.120				.886	.157
Planning				.072	.034	.158*		
WM				.026	.012	.169*		
LTM				.014	.005	.203**		
Paso 4	9.039** (4163)	.182	.161				.973	.222
Planning				.053	.034	.115		
WM				.022	.012	.146*		
LTM				.013	.005	.191**		
VP				.018	.006	.220**		

*Note: WS: Writing Sample, WF: Writing Fluency, WE: Written Expression; VP: Visuoperception; WM: Working Memory; LTM: Long-Term Memory. * < .05; ** < .01.*

Table 1.
Hierarchical regression of Planning, Working Memory, Long-Term Memory and Visuoperception on Written Expression.

It is worth noting that working and long-term memory, along with visuo-spatial skill, were the main predictors for expressive writing skills according to the regression model. It has been claimed that working memory is engaged for the integration of lexical-semantic and syntax processes in sentence building during writing [7, 23, 42, 54, 55], and it has been shown that children with higher working memory performances produced texts of better quality and syntactic complexity [41, 45]. Furthermore, it has been proposed that working memory is directly related to the text generation component of the translating process [56]. In turn, working memory provides a workspace to manipulate those contents retrieved from long term memory systems, which include vocabulary, grammar, syntax and orthographic rules involved in sentence generation [23]. Both memory systems interact and collaborate to allow and facilitate lower-level writing processes, by easing the access and manipulation of the linguistic information required to build phrases and sentences. In addition, mature visuospatial skills are required for learning to write, and they are engaged in the automatization of orthographic coding and graphomotor processes [47, 56]. It is then expected that more fluent and efficient writing is linked to better visuospatial skills among children [57].

In sum, Moreno's study [24] indicated that working memory, long-term memory retrieval and visuospatial skills contribute to children's low-level translating processes to achieve coherent and fluent sentence writing. In the following section, we will refer to a study that examined the role of these memory systems in children's higher-level writing skills, analyzing different text dimensions as the outcome of planning and reviewing processing.

3.2 The role of long-term memory and working memory in higher level writing processes

The following section addresses the link between higher-level writing processes and children's working and long-term memory systems. While lower-level aspects of writing (such as transcription) have been more studied, there are relatively few studies of planning and reviewing processes [23, 37, 58], and most of them have focused on secondary and university students [59, 60]. A recent study that examined writing planning processes among 1558 Argentinean primary school children found that the length, quality and complexity of their narrative texts increased as a function of grade, which indicated an effect of schooling, practice and cognitive development [37]. Previous research further indicated that planning and reviewing skills are not fully operational during primary school, and only contribute to text quality in older children [61].

As we stated before, *planning* [25, 62] involves setting goals, retrieving and organizing relevant contents from long-term memory and facing rhetorical problems. *Reviewing* requires scanning the text for errors, evaluating if it meets writing goals and revising when necessary. The quality of written texts can be examined by assessing a range of dimensions. The microstructure refers to local meaning and includes within and between-propositional order and coherence. The macrostructure refers to global meaning and encompasses communicative purposes (pragmatics) and message topics (semantics). These macrostructures can be recovered from long-term memory to aid text building processes [63, 64].

A recent study by Moreno [7] examined the association between school-aged children working and long-term memory system's performance and the quality of their narrative texts, considering micro and macrostructural dimensions. A total of 83 9–11-year-old children, from the 4th and 5th primary school grades participated in the study. Children came from medium and low socioeconomic backgrounds, and in some of them lived in socially vulnerable environments. The children completed the following battery of psycholinguistic and neuropsychological tests:

- Working memory and long-term memory retrieval were assessed using the Auditory Working Memory and the Retrieval Fluency subtests from the Woodcock-Johnson III battery [65].

- Text quality: Children were asked to write narrative texts in response to three pictures (a fantasy scene, a series of actions and a daily life scene). An ad hoc instrument was designed to evaluate the quality of the texts texts along the following dimensions: *pragmatics* (adequacy of text purpose and intended audience), *superstructure* (adequacy of canonic categories and schema of narrative plot), *macrostructure* (global semantic content unity and coherence), *microstructure* (adequacy of thematic progression and cohesion between sentences), *propositional* (internal structure within sentences) and *orthographic* (following textualization conventions and rules). The instrument showed adequate psychometric properties [7].

The quality of the children's narrative texts was categorized for each dimension as low, medium or high, according to their respective scores. The best performances were observed in the *pragmatics* dimension, where most children obtained medium (48.3%) or high (50.6%) scores. On the other hand, *superstructure* exhibited the highest proportion of low scores (87.4%). Performance was mostly medium in the *macrostructure* (47.1%) and mostly low in *microstructure* (77%). *Medium* scores were the most frequent in the propositional (90.8%) and orthographic (71.3%) dimensions. Therefore, children's performance was medium or low in 5 of the 6 dimensions, exhibiting greater difficulties in the superstructure and the microstructure of the text. In addition, significant and moderate correlations were observed among all dimension scores ($.296 < rho < .808$, p's $< .01$), indicating that those students who achieved a higher quality in pragmatic and contextual aspects of composition also performed well at the structural and syntactic levels.

Regarding the associations with memory systems, significant correlations were found between working memory and: pragmatics, superstructure, macrostructure, and orthographic dimensions ($.232 < rho < .363$, p's $< .01$); and between long-term memory and: pragmatics, macrostructure, propositional and orthographic dimensions ($.224 < rho < .312$, p's $< .01$). To further examine these relations, each dimension was analyzed with an ANOVA, considering working memory and long-term memory performance as the predictor (categorized as low, medium or high) (see **Table 2**).

Children with high working memory performance exhibited better scores than those with low performance in the pragmatic and macrostructural dimensions (p's $< .033$), while this effect was marginally significant for superstructure ($p = .053$) and orthographic ($p = .056$) dimensions. In turn, the high long-term memory group outperformed the low group in the pragmatic and macrostructure dimensions (p's $< .022$), while this effect was marginally significant for superstructure ($p = .054$). These results indicate that the major contribution of working and long-term memory systems was observed in the macrostructural dimensions of the text, including aspects such as communicative purposes topic adequacy, narrative structure and global cohesion.

Moreno study [7] highlights the contribution of children's working and long-term memory systems to pragmatic and macrostructural aspects of their written narratives, which can be understood in terms of their engagement in planning and reviewing processes. Working memory provides a workspace for the selection and manipulation of relevant information while maintaining the writing goals online [26]. Vocabulary, topic and world knowledge can be chosen to match the intended readers' profile. Working memory also allows integrating the theme, characters, narrator, time and space information into a cohesive narrative, as well as the spatial diagraming of the text. Furthermore, working memory is engaged by text generation in the translating process, and during evaluating and revising in the reviewing process. It is worth noting that the theorized role of working memory in writing processes bears resemblance to its theoretical contribution to reading comprehension [3], where it supports the integration of long-term memory contents activated by text input to generate local and global representations of the text.

Long-term memory retrieval is acknowledged as a fundamental step in the planning processes, allowing episodic, semantic and linguistic knowledge into the working memory [66, 67]. In addition, textual macrostructures rely on episodic and semantic memories to give shape to discourse, and they are stored as templates in long-term memory. These templates are brought into working memory to drive and organize text planning and generation processes [63, 64, 68].

We should point out that, despite the predominance of medium and low-quality scores among children's narratives, Moreno [7] could still observe a significant contribution of memory systems to pragmatic and macrostructural dimensions. This suggests that both working memory capacity and long-term retrieval processes

Dimension/ Working Memory	M	DS	F	gl	p*	Dimension/ Long Term Memory	M	DS	F	gl	p*
Pragmatics	1.91	.765	3.710	2	.030	Pragmatics	2.02	.73	3.567	2	.022
Low-WM (n = 19)	2.32	.773				*Low-LTM (n = 22)*	2.22	.76			
Medium-WM (n = 45)	2.60	.632				*Medium-LTM (n = 40)*	2.71	.56			
High-WM (n = 19)						*High-MLP (n = 21)*					
Superstructure	.670 .84	.970	2.756	2	.071	Superstructure	.41	.59	2.957	2	.054
Low-WM (n = 19)	1.73	1.405				*Low-LTM (n = 22)*	1.22	1.56			
Medium-WM (n = 45)		2831				*Medium-LTM (n = 40)*	1.24	1.44			
High-WM (n = 19)						*High-MLP (n = 21)*					
Macrostructure	4.61	2.320	6.338	2	.003	Macrostructure	5.13	2.41	7.091	2	.018
Low-WM (n = 19)	5.95	2.092				*Low-LTM (n = 22)*	5.89	1.88			
Medium-WM (n = 45)	7.13	1.506				*Medium-LTM (n = 40)*	6.95	1.98			
High-WM (n = 19)						*High-MLP (n = 21)*					
Microstructure	.83	1.042	2.133	2	.126	Microstructure	.49	.96	1.632	2	.282
Low-WM (n = 19)	.81	1.204				*Low-LTM (n = 22)*	0.92	1.40			
Medium-WM (n = 45)	1.60	1.765				*Medium-LTM (n = 40)*	1.10	1.34			
High-WM (n = 19)						*High-MLP (n = 21)*					
Propositional	5.95	1.590	2.944	2	.059	Propositional	5.46	1.40	2.485	2	.145
Low-WM (n = 19)	6.00 7.13	1.452				*Low-LTM (n = 22)*	6.23	1.79			
Medium-WM (n = 45)		2.066				*Medium-LTM (n = 40)*	6.43	1.94			
High-WM (n = 19)						*High-MLP (n = 21)*					
Orthographic	7.97	4.661	2.987	2	.057	Orthographic	8.93	3.86	2.985	2	.121
Low-WM (n = 19)	8.42	3.475				*Low-LTM (n = 22)*	9.33	4.02			
Medium-WM (n = 45)	11.0	3.873				*Medium-LTM (n = 40)*	8.55	3.42			
High-WM (n = 19)						*High-MLP (n = 21)*					

$^{*}p < 0{,}05$.

Table 2.
Descriptive statistics and Anovas of writing dimensions, according to the groups of low, medium and high performance in Working Memory and Long Term Memory.

allow to manage and optimize the children's cognitive resources to achieve thematic adequacy and semantic coherence.

4. Conclusions

Throughout this chapter, we have outlined the theoretical role of working and long-term memory systems in producing written text, explained the relevance of the ontogenetic trajectories of frontoparietal and MTL cortexes for the development of these memory systems and discussed the empirical evidence of their contribution to children's high and low-level writing skills. We found that long-term memory retrieval and information integration within working memory are significantly linked to the capacity of conceiving a text, planning its redaction and solving rhetorical problems, but are also involved in the ideation and transcription processes that translate children's thoughts into handwritten words. We elaborate on these findings below.

The positive correlation of memory systems' performance with most of the structural dimensions of the written texts shows that working and long-term memory contribute to the quality of the texts. These findings are congruent with previous research indicating the importance of memory systems to integrate new learning with previously acquired information, using it to write and solve problems [69–71].

Duff and Brown-Schmidt' [32] proposed that the hippocampal declarative memory system contributes to online language processing given its capacity for maintenance and integration of multimodal representations. In Hasson et al. [72], the medial temporal hippocampal region, would interact with long-term memory regions, facilitating the consolidation of incoming information with the global context and prior knowledge of the world. While Duff & Brown-Schmidt [32] hypothesis referred to speech comprehension and production, it could also be extended to interpret findings from reading and writing studies. In this way, it seems that the supplementary motor area (SMA) and the hippocampus play a fundamental role in reading comprehension, as was shown in the study by Hsu et al. [34], since the predictive and integrative processes unfold independently of the textual genre (narrative or expository). In the same line, Shah et al. showed how hippocampal activation contributed to memory retrieval and semantic integration during creative writing tasks.

Learning to write requires accessing conceptual networks mapped and stored within semantic memory. Procedural memories of syntax and orthographic rules are also engaged. In addition, the intervention of working memory allows choosing and organizing the necessary information to build the text and coordinating cognitive resources to write it [1, 7, 66, 70]. This coordination allows children to display a series of microprocesses that come together in the composition process. That is, they can accurately select the topic and structure according to the text genre, plan its spatial layout, situate the actions in time and space, and place a narrator [7].

In addition, long-term retrieval of declarative, semantic and linguistic expression knowledge, allows to fulfill the pragmatic and macrostructural levels of writing [69, 70]. Along with storage capacity, the ability to retrieve that information efficiently aids the text planning processes. This process involves the deployment of other skills such as: identification of facts or concepts, associative thinking and expressive fluency. These processes allow the effective transfer of enunciative knowledge and procedures to immediate consciousness, through its connection with working memory [66–68].

On the other hand, those models focusing on textual production emphasize that textual macrostructures organize themselves to configure a coherent text. In this organization, episodic and semantic memories provide the necessary information, as they are based on life experiences and stored conceptual knowledge. Consequently, textual macrostructures are stored in long-term memory and work as templates that, when retrieved and organized in working memory, guide the rest of the textual production processes [63, 68, 69].

As a conclusion, we can say that the contribution of memory performance to children's writing skills, shows the importance of stimulating the development of these systems within the school environment. This stimulation could be a relevant factor in learning to write, visuospatial and motor skills, as well as in the fluency of ideas necessary for an adequate quality of the text. The link between writing and long-term memory is further supported by another recent study [33] that found that a handwriting task activated parietal and central brain areas in young adults and (to a lesser extent), 12-year-old children. Neural activity in these areas is important for memory and for the encoding of new information, thus providing the brain with optimal conditions for learning. Therefore, the authors suggested that early exposure to writing at school would help to establish brain oscillation patterns that are beneficial for learning. In addition, sensorimotor integration along with fine, controlled hand movements for writing are vital to facilitating and sustaining learning. These conclusions are further supported by converging evidence that shows how the practice of expressive writing can have beneficial effects on children and adults' working [73–75] and long-term memory [76] performance, therefore highlighting the link between memory systems and writing skills.

Author details

Cecilia Beatriz Moreno* and Ángel Javier Tabullo
Pontifical Catholic University of Argentina, Institute of Social Environmental and Human Sciences, Mendoza, Argentina

*Address all correspondence to: cecilia_moreno@uca.edu.ar

References

[1] Ferreres A, China N, Abusamra V. Cerebro, desarrollo y educación. In: J. J. A. Castorina y M. Carretero, compiladores, editor. Desarrollo Cognitivo y Educación. Los Inicios del Conocimiento. Buenos Aires: Paidós; 2012. pp. 113-131

[2] Duke NK, Cartwright KB. The science of reading progresses: communicating advances beyond the simple view of reading. Reading Research Quarterly. 2018;**56**:25. DOI: 10.1002/rrq.411

[3] Butterfuss R, Kendeou P. The role of executive functions in reading comprehension. Educational Psychology Review. 2018;**2018**(30):801-826

[4] Berninger VW, Winn WD. Implications of advancements in brain research and technology for writing development, writing instruction, and educational evolution. In: MacArthur C, Graham S, Fitzgerald J, editors. Handbook of Writing Research. New York: Guilford; 2006. pp. 96-114

[5] Kim YG, Schatschneider C. Expanding the developmental models of writing: A direct and indirect effects model of developmental writing (DIEW). Journal of Educational Psychology. 2017;**109**(1):35-50. DOI: 10.1037/edu0000129

[6] Lacón N, Ortega S. Producción de Textos Escritos. Mendoza: EDIUNC; 2013

[7] Moreno C. Escritura de textos narrativos en niños escolares: la importancia de la memoria de trabajo y la memoria a largo plazo. Revista de Psicología. 2016;**12**(24):7-17

[8] Cassandra B, Reynolds C. A model of development of frontal lobe functioning: Findigs from a Meta-analysis. Applied Neuropsychology. 2005;**12**(4):190-201. DOI: 10.1207/s15324826an1204_2

[9] Schneider W, Ornstein PA. Determinants of memory development in childhood and adolescence. International Journal of Psychology. 2018;**54**(3):307-315. DOI: 10.1002/ijop.12503

[10] De Haan M, Mishkin M, Baldeweg T, Vargha-Khadem F. Human memory development and its dysfunction after early hippocampal injury. Trends in Neurosciences. 2006;**2006**(29):374-381. DOI: 10.1016/j.tins.2006.05.008

[11] Shing Y, Werkle-Bergner M, Brehmer Y, Müller V, Li S, Lindenberger U. Episodic memory across the lifespan: The constibutions of associative and estrategia componente. Neuroscience & Biobihavioral Reviews. 2010;**34**(7):1080-1091. DOI: 10.1016/j.neubiorev.2009.11.002

[12] Johnson M, de Haan M. Developmental Cognitive Neuroscience: An Introduction. Fourth ed. West Sussex, UK: John Wiley & Sons, Ltd; 2015. Available from: https://www.wiley.com/go/johnson/developmentalcognitiveneuroscience

[13] Bauer PJ, Dugan JA. Memory, development. In Rubinstein JP. Rakic, editors. Neural Circuit and Cognitive Development, 2nd Edition. London: Academic Press, Elsevier; 2020. pp. 395-412

[14] Cowan N. Short-term and working memory in child- hood. In: Bauer PJ, Fivush R, editors. The Wiley Hand-Book on the Development of children's Memory. Vol. II. Chichester, U.K: Wiley-Blackwell; 2014. pp. 202-229

[15] Klingberg T, Forssberg H, Westerberg H. Increased brain activity in frontal and parietal cortex underlies the development of visuospatial working memory capacity during childhood. Journal of Cognitive Neuroscience. 2002;**14**(1):1-10. DOI: 10.1162/089892902317205276

[16] Barrouillet P, Camos V. The time-based resource-sharing model of working memory. In: Osaka N, Logie RH, D'Esposito M, editors. The Cognitive Neuroscience of Working Memory. Oxford, U.K: Oxford University Press; 2007. pp. 59-80

[17] Tamnes CK, Walhovd KB, Grydeland H, Holland D, Østby Y, Dale AM, et al. Longitudinal working memory development is related to structural maturation of frontal and parietal cortices. Journal of Cognitive Neuroscience. 2013;**25**(10):1611-1623. DOI: 10.1162/jocn_a_00434

[18] Darki F, Klingberg T. The role of Fronto-parietal and Fronto-striatal networks in the development of working memory: A longitudinal study. Cerebral Cortex. 2015;**25**(6):1587-1595. DOI: 10.1093/cercor/bht352

[19] Satterthwaite TD, Wolf DH, Erus G, Ruparel K, Elliott MA, Gennatas ED, et al. Functional maturation of the executive system during adolescence. Journal of Neuroscience. 2013;**33**(41): 16249-16261. DOI: 10.1523/JNEUROSCI.2345-13.2013

[20] Rosenberg MD, Martinez SA, Rapuano KM, Conley MI, Cohen AO, Cornejo MD, et al. Behavioral and neural signatures of working memory in childhood. The Journal of Neuroscience: The Official journal of the Society for Neuroscience. 2020;**40**(26):5090-5104. DOI: 10.1523/JNEUROSCI.2841-19.2020

[21] Kail RV, Ferrer E. Processing speed in childhood and adolescence: Longitudinal models for examining developmental change. Child Development. 2007;**78**(6):1760-1770. DOI: 10.1111/j.1467-8624.2007.01088.x

[22] Camos V, Barrouillet P. Developmental change in working memory strategies: From passive maintenance to active refreshing. Developmental Psychology. 2011;**47**(3):898-904. DOI: 10.1037/a0023193

[23] Abusamra V, Miranda A, Cartoceti R, Difalsis M, Re A, Cornoldi C. Batería para la evaluación de la escritura, 2020. Buenos Aires: Paidós; 2020

[24] Moreno CB, Korzeniowski C, Espósito A. Cognitive and executive processes associated with children's writing. OCNOS. Journal of Reading Research. 2022;**21**:2. DOI: 10.18239/ocnos_2022.21.2.2839

[25] Flower L, Hayes J. A cognitive process theory of writing. College Composition and Communication. 1981;**32**(4):365-387. DOI: 10.2307/356600

[26] Olive T. Writing and working memory: A summary of theories and findings. In: Grigorenko EL, Mambrino E, Preiss DD, editors. Writing: A Mosaic of New Perspectives. New York: Psychology Press; 2012. pp. 125-140

[27] Miyake A. Individual differences in working memory: Introduction to the special section. Journal of Experimental Psychology: General. 2001;**130**(2):163-168. DOI: 10.1037/0096-3445.130.2.163

[28] Drijbooms E, Groen MA, Verhoeven L. The contribution of executive functions to narrative writing in fourth grade children. Reading and Writing. 2015;**28**(7):989-1011. DOI: 10.1007/s11145-015-9558-z

[29] Salas N, Silvente S. The role of executive functions and transcription skills in writing: A cross-sectional study across 7 years of schooling. Reading and Writing. 2019;**33**:877-905. DOI: 10.1007/s11145-019-09979-y

[30] Bryson M, Bereiter C, Scardamalia M, Joram E. Going beyond the problem as given: Problem solving in expert and novice writers. In: Stumberg R, Frensch P, editors. Complex Problem Solving. Hillsdale, NJ: Lawrence Erlbaum; 1991

[31] Bereiter C, Scardamalia M. The Psychology of Written Composition. Hillsdale, NJ: Erlbaum; 1987

[32] Duff MC, Brown-Schmidt S. The hippocampus and the flexible use and processing of language. Frontiers in Human Neuroscience. 2012;**2012**:6. DOI: 10.3389/fnhum.2012.00069

[33] Shah C, Erhard K, Ortheil HJ, Kaza E, Kessler C, Lotze M. Neural correlates of creative writing: An fMRI study. Human Brain Mapping. 2013;**34**(5):1088. DOI: 10.1002/hbm.21493

[34] Hsu C, Clariana R, Li P. Neurocognitive Signaturas of naturalistas reading of scientific texts: A fixatio-related fMRI study. Intertantional Journal for Equity Health. 2020;**9**:10678. DOI: 10.1038/s41598-019-47176-7

[35] Ose AE, van der Weel FR, van der Meer ALH. The importance of cursive handwriting over typewriting for learning in the classroom: A high-density EEG study of 12-year-old children and young adults. Frontiers in Psychology. 2020;**11**:1810. DOI: 10.3389/fpsyg.2020.01810

[36] van der Meer ALH, van der Weel FRR. Only three fingers write, but the whole brain works: A high-density EEG study showing advantages of drawing over typing for learning. Frontiers in Psychology. 2017;**8**:706. DOI: 10.3389/fpsyg.2017.00706

[37] Difalcis M, Cartoceti R, Abusamra V. Procesos de composición escrita en la escuela primaria: un estudio relegado. Traslaciones. Revista Latinoamericana De Lectura Y Escritura. 2022;**9**(17):196-217. DOI: 10.48162.rev.5.063

[38] Ministerio de Educación de la Nación Argentina. Argentina en PISA 2018. Informe de resultados. 2018. Available from: https://www.argentina.gob.ar/educacion/evaluacion-informacion-educativa/ internacionales/pisa

[39] Ministerio de Educación de la Nación Argentina. Aprender 2017. Primaria: Informe de Resultados; 2017. Available from: https://www.argentina.gob.ar/educacion/aprender2017/reportes-nacionales

[40] Berninger VW, Yates CM, Cartwright AC, Rutberg J, Remy E, Abbott RD. Lower-level developmental skills in beginning writing. Reading and Writing: An Interdisciplinary Journal. 1992;**4**:257-280. DOI: 10.1007/BF01027151

[41] Cordeiro C, Limpo T, Olive T, Castro S. Do executive functions contribute to writing quality in beginning writers? A longitudinal study with second graders. Reading and Writing. 2019;**2019**(33):813-833. DOI: 10.1007/s11145-019-09963-6

[42] Miranda A. Desarrollo de la escritura en español: un estudio psicolingüístico. Buenos Aires: Universidad de Buenos Aires; 2019

[43] Alves RA, Limpo T. Progress in written language bursts, pauses, transcription, and written composition across schooling. Scientific Studies of Reading. 2015;**19**(5):374-391. DOI: 10.1080/10888438.2015.1059838

[44] Hooper SR, Swartz C, Wakley MB, de Kuif RE, Montgomery J. Executive functions in elementary school children with an without problems in written expression. Journal of Learning Disabilities. 2002;**35**:57-68. DOI: 10.1177/002221940203500105

[45] Drijbooms E, Groen M, Verhoeven L. How executive functions predict development in syntactic complexity of narrative writing in the upper elementary grades. Read and Writing. 2017;**30**:209-231. DOI: 10.1007/s11145-016-9670-8.pdf

[46] Altemeier L, Jones J, Abbott R, Berninger V. Executive functions in becoming Writting readers and readers writers: Note taking and report writing in third and fifth grades. Developmental Neuropsychology. 2006;**29**(1):161-173. DOI: 10.1207/ s15326942dn2901_8

[47] Bravo-Cóppola L. Las destrezas perceptuales y los retos en el aprendizaje de la lectura y la escritura. Una guía para la exploración y comprensión de dificultades específicas. Actualidades Investigativas en Educación. 2004;**4**:1. DOI: 10.15517/aie.v4i1.9047

[48] Olive T, Passerault JM. The visuospatial dimension of writing. Written Communication. 2012;**29**(3):326-334. DOI: 10.1177/0741088312451111

[49] García-Pérez EM, Magaz-Lagos A. Escalas Magallanes de Atención Visual. Madrid: Albor- Cohs; 2000

[50] Carrada M, Ison M. La eficacia atencional. Un estudio normativo en niños escolarizados de Mendoza. PSIENCIA. Revista latinoamericana de Ciencia Psicológica. 2013;**5**(2):63-73. DOI: 10.5872/psiencia/5.2.22

[51] Marino J, Fernández A, Alderete A. Valores Normativos y Validez Conceptual del Test de Laberintos de Porteus en una Muestra de Adultos Argentinos. Revista Neurológica Argentina. 2001;**26**:102-107

[52] Espósito A, Ison M. Aportes a los criterios de evaluación de la Figura Geométrica Compleja de Rey (Figura A). Acta Psiquiátrica y Psicológica de América Latina. 2011;**57**(4):299-307

[53] Wendling BJ, Schrank FA, Schmitt AJ. Educational Interventions Related to the Woodcock-Johnson III Tests of Achievement, Bulletin No. 8. Itasca, IL: Riverside Publishing; 2007

[54] Cuetos F. Psicología de la Escritura. Madrid: Editorial Escuela Española; 2009

[55] Hayes JR. Kinds of knowledge-telling: Modeling early writing development. Journal of Writing Research. 2011;**3**(2):73-92. DOI: 10.17239/jowr-2011.03.02.1

[56] Berninger VW. Coordinating transcription and text generation in working memory during composing: Automatic and constructive processes. Learning Disability. 1999;**22**(2):99-112. DOI: 10.2307/1511269

[57] Olive T, Kellogg RT, Piolat A. Verbal, visual and spatial demands during text composition. Applied Psycholinguistics. 2008;**29**(4):669-687. DOI: 10.1017/ S0142716408080284

[58] Tressoldi PE, Cornoldi C, Re AM. Batteria per la Valuatzione della Scrittura e della Competenza Ortografica 2 (BVSCO-2). Roma: IGuinti Organizzazioni Speciali; 2012

[59] Dessau R. Escribir en la Universidad: Un desafío de creatividad y pensamiento crítico. Paidós: Buenos Aires; 2016

[60] Navarro F, Revelchion A. Escribir para aprender: disciplinas y escritura en la escuela secundaria. Paidós: Buenos Aires; 2013

[61] Limpo T, Alves R, Fidalgo R. Children's high-level writing skills: Development of planning and revising and their contribution to writing quality. British Journal of Educational Psychology. 2014;**84**(2):177-193. DOI: 10.1111/bjep.12020

[62] Hayes JR. A new framework for understanding cognition and affect in writing. In: Levy CM, Ransdell SR, editors. The Science of Writing. Mahwah, NJ: Lawrence Erlbaum Associates; 1996. pp. 1-27

[63] Molina García S. Psicopedagogía de la Lengua Escrita: Lectura. Madrid: Editorial EOS; 2008

[64] Van Dijk TA, Kinstch W. Strategies for Discourse Comprehension. New York: Academic Press; 1998

[65] Muñoz-Sandoval AF, Woodkock RW, Mc Grez KS, Mather N. Batería III Woodkock- Muñoz. Itasca, IL: Riverside Publishing; 2005

[66] Baddeley A. Exploring the central executive. The Quarterly Journal of Experimental Psychology Section A. 1996;**49**:5-28. DOI: 10.1080/713755608

[67] García Madruga J, Gutiérrez Martínez F, Vila Chaves J. Desarrollo Cognitivo y Educación II. Procesos del Conocimiento y Contenidos Específicos. Vol. 2012. Paidós: Buenos Aires; 2012. pp. 47-69

[68] Mather N, Woodcock R, Manual del examinador, L. Wolfson, Traductor. Woodcock- Johnson III Pruebas de Habilidades Cognitivas. Itasca, IL: Riverside Publishing; 2005

[69] Van Dijk TA. Ideología y Análisis del Discurso. Utopía y Praxis Latinoamericana. 2005;**10**(29):9-36. Available from: http://ve.scielo.org/scielo.php?script=sci_arttext&pid=S1315-52162005000200002

[70] Montoya D, Varela V, Dussan C. Correlación entre las habilidades académicas de lectura y escritura y el desempeño neuropsicológico en una muestra de niños y niñas con TDAH de la ciudad de Manizales. Psicología desde el Caribe. 2012;**29**(2):305-329. Available from: https://rcientificas.uninorte.edu.co/index.php/psicologia/article/view/3399

[71] Tulving E. Memory. In: Gazzaniga M, editor. The New Cognitive Neuroscience. Cambridge, MA: MIT Press; 2000. pp. 727-733

[72] Hasson U, Chen J, Honey CJ. Hierarchical process memory: Memory as an integral component of information processing. Trends in Cognitive Science. 2015;**19**(6):304-313. DOI: 10.1016/j.tics.2015.04.006

[73] Fuster J. The prefrontal córtex- an update: Time of the essence. Neuron. 2001;**30**(2):319-333. DOI: 10.1016/S0896-6273(01)00285-9

[74] Fartoukh M, Chanquoy L. Expressive writing in school children: Effects on well-being and working memory. Journal of Writing Research. 2020;**11**(3):505-523. DOI: 10.17239/jowr-2020.11.03.04

[75] Klein K, Boals A. Expressive writing can increase working memory capacity. Journal of Experimental Psychology: General. 2001;**130**(3):520-533. DOI: 10.1037/0096-3445.130.3.520

[76] Conway MA, Gathercole SE. Writing and long-term memory: Evidence for a "translation" hypothesis. The Quarterly Journal of Experimental Psychology Section A. 1990;**42**(3):513-527. DOI: 10.1080/14640749008401235

Chapter 5

Performing Music on Stage: The Role of the Hippocampus in Expert Memory and Culture

Christiane Neuhaus

Abstract

This overview chapter discusses memory functions from the viewpoint of the performing arts. 'Playing music by heart' is taken as an example to illustrate the role of the hippocampus in acquiring and expressing expert memory. Many more aspects depend on hippocampal mechanisms beyond declarative memory, for example, motor sequence learning, phrase boundary processing, and time-precise sequence recall. In consequence, changes in size and/or functional activity also occur in the hippocampus, known as hippocampal plasticity. Whenever the to-be-remembered items have to be stabilized even further, certain mnemonic strategies are effective, of which the oldest is the (hippocampal-based) method of loci, using visuospatial imagery. Mnemonic techniques also play a role in ethnomusicology. For example, North Indian tabla players combine drum patterns with certain onomatopoeic syllables to keep on track when performances last over hours. The value of memory processes is also discussed from a sociocultural perspective. Since priests, teachers, heads of tribes, and many others are explicit carriers of internalized knowledge, they help preserve oral traditions and culture. A special emphasis is on the accurate memorization of the Quran in Arabic, revealing that internalized sacred knowledge acquired through learning by rote can serve as a moral compass for the individual.

Keywords: hippocampus, expert memory, neuroplasticity, recognition memory, chunking, mnemonic techniques, method of loci, North Indian tabla playing, Quran memorization, culture

1. Introduction

Soloists sometimes produce 1800 tones per minute when performing live on stage. Note-perfect, virtuoso recitals have their roots in the nineteenth century, in the works by Franz Liszt, Paganini, and Frédéric Chopin. By convention, and for better expressiveness, skillful pieces are played by heart. For that, expert memory [1] is a precondition manifesting itself in time-precise and highly accurate sequence recall. This way every structural detail (and nuance) of the piece can be retrieved in a stable manner, also when performances last 20 min or longer (**Figure 1**). Expert memory, therefore, differs in many respects from self-paced (episodic) recall, that is, from the standard type of sequence memory as in storytelling.

Figure 1.
Expert memory is a precondition for expressive performance without the score.

It seems reasonable to assume that hippocampal mechanisms are involved, both during memorizing and recall. However, in the field of "Music and Neuroscience" (also termed "Neuromusicology") studies on the relationship between (active) long-term memory processes and hippocampal structures are scarce. The most crucial question in this respect is if the human hippocampus is sensitive to temporal details at all, in particular as regards tone lengths, micro-pauses, and the sequence order of elements. A preliminary proof has already been given in the visual domain: Barnett et al. [2] could show in a functional MRI study, using a match/mismatch detection paradigm and picture series as stimuli, that both hippocampi were strongly activated when the temporal information was held constant, that is, when event lengths and inter-stimulus durations within the picture series were exactly the same at encoding and the test phase. Note that this is an example of retrieval success (see, e.g., [3]). However, closer to a musician's reality is an fMRI study from the field of psycho-acoustics [4] in which hippocampal sensitivity to temporal details was put to the test once again: In this study [4], a series of monotonous but otherwise regularly spaced pure tones (offset-onset distance: 150 ms) was occasionally interspersed with pure tones occurring earlier, that is, with offset-onset intervals of 142-, 130-, and 100-ms duration, resulting in a sort of "stumbling" listening impression. (Note that detecting these too-early-appearing deviants was more or less automatic since participants were instructed to ignore the sounds while watching a silent movie.) Sections of temporal irregularity activated the left anterior hippocampus in a group of professional musicians, which was not the case for the group with nonmusician controls. (However, for both subject groups, an enhanced BOLD response could be observed in the right planum temporale.) Note that in [4], hippocampal activity was induced by "novelty in temporal structure," whereas in the previous study with picture series [2] the hippocampus reacted to sameness, since all temporal details were kept constant.

This inconsistency in the results might be attributed to the fact that [4] had all the hallmarks of a classical mismatch design, whereas Barnett et al. [2] did not. Some scientists also point out that the cell types within the hippocampus may have different properties, with one type of neurons showing reactions to novelty, whereas the other showing reactions to items remembered as old. So, all in all, the human hippocampus indeed seems to be sensitive to fine-grained temporal aspects, which is a precondition for building exact representations of musical sequences in mind and making the sound results stable. (To some degree this holds also for actors learning their roles).

However, with these findings a basic problem becomes obvious: Gaps in research, especially missing functional MRI data on how the hippocampus might react to musical features other than those related to time, make it unavoidable to sometimes argue by analogy. In other words, conclusions occasionally have to be drawn from results obtained in other domains, outside the field of neuromusicology. In addition, it would be more trustworthy if results would be less contradictory as is the case with [4] compared with [2]. Thus, to achieve homogeneity more studies are needed in the field of Music and Neuroscience with special focus on hippocampal activity (replication studies included).

On the other hand, a considerable number of hippocampal studies deal with episodic memory, that is, when certain personal memories are built or retrieved. The respective key terms are: "binding," "linking," or "relating" persons and/or disparate objects with each other. Furthermore, to achieve coherence persons and/or objects have to be set into the situational context, that is, events must be adjusted to space and time (e.g., [5, 6]). In this regard, Maguire and Mullally [7] recently put emphasis on the words "scene" and "scene construction" underlining that episodic memory is holistic (or: pattern-like) in nature. Accordingly, episodic recall is completely different and has to be distinguished from time-precise sequence recall needed to play musical pieces by heart. Thus, drawing parallels between both memory types has to be considered with caution.

This book chapter discusses memory functions from the viewpoints of the performing arts and culture. Wherever possible, the focus is on hippocampal activity in terms of processing music. The text is divided into three parts. The first gives an overview of studies from the field of Music and Neuroscience, in particular, what has been found out about hippocampal plasticity and hippocampal involvement regarding recognition memory. The second part starts from the observation that soloists work hard to make the recall of musical passages during performance more stable. It will become evident that a distinguishing feature of expert memory is the use of certain mnemonic strategies, including a grouping mechanism called "chunking." Maguire et al. [8] found out by questionnaire that superior memorizers instinctively (but consciously) use mnemonic strategies when asked to memorize some simple sequences, whereas matched control subjects do not. A special focus is on the so-called method of loci, an old mnemonic strategy already known by the ancient Greeks and Romans. Interestingly, a special case of mnemonic technique has been discovered in the field of ethnomusicology: While playing drum patterns North Indian tabla players think in mnemonic syllables and say them aloud to keep on track when performances last over hours. Finally, part three goes beyond individual memory, showing the role of mnemonic devices in preserving oral traditions and culture. A special emphasis is on the accurate memorization of the Quran.

First, I will provide some background on the anatomical facts and findings that are relevant for this chapter*: The anatomical core regions necessary to bring thoughts, facts, and events back to mind are the hippocampus proper and its adjacent structures located in the medial temporal lobe (MTL), the so-called perirhinal, parahippocampal, and entorhinal cortices. On the whole, highly pre-processed information reaches the hippocampus proper, which is often described as a "convergence zone" (e.g., [6, 9], see also [10]). Today, the conventional way of assigning memory functions to the left and the right hippocampus has been given up. According to this oversimplified view, the left hippocampus is involved in the storage of personal events (or: episodic memories), whereas the right is involved in spatial navigation and the storage of places (for this see, e.g., [11]). At present, the focus has shifted towards the long axis of the hippocampus, and specific functions are associated with its anterior and posterior subregions [12]. Note that some studies use both reference systems, long-axis as well as laterality (e.g., [13]). There is strong agreement that spatial information is processed in the posterior subregions bilaterally, whereas semantic aspects mainly activate the left anterior part. (The latter points to the substantial role the hippocampus has in terms of both the semantic and the episodic types of declarative memory.) Moreover, based on progress in measurement technology, the focus has shifted further towards anatomical and functional connectivity, which is investigated with diffusion tensor imaging (DTI, e.g., [14], see also [9]). Accordingly, the hippocampus is no longer considered as an isolated structure but is rather seen in conjunction with neocortical and subcortical brain regions, to those it is linked via strong reciprocal pathways. Thus, hippocampal research now pursues a more dynamic approach, and this has been extended even further: Ritchey et al. [15] propose the existence of two large-scale systems interacting with each other, a posterior medial system (PM) and an anterior-temporal one (AT) subsumed under the acronym "PMAT framework." Due to its strong connectivity with both systems the hippocampus plays dual roles in PMAT and serves also as an integration zone.

Reciprocal (or: bidirectional) connections between the hippocampus and the neocortex also play a decisive role in regard to memory consolidation, which is often described as "resistance to interference [16]." It is common knowledge that the hippocampus merely acts as a temporary store. That is, even though the memorized content initially depends on hippocampal processes to make it resistant to interference on the molecular level, it is passed on to neocortical regions to be stored there in a distributed but permanent manner. Although there is strong consensus on these hippocampo-cortical mechanisms regarding "the way to get there," that is, while building stable memory traces, researchers disagree on the reverse, that is, on the role the hippocampus has when stored content is re-activated during retrieval. Some researchers are convinced that long-term information is retrieved independently of hippocampal activity, since, approximately 24 h after encoding, cortico-cortical connectivity increases, while hippocampo-cortical connectivity decreases (see, e.g., [17] for details). In contrast to that, Teyler et al. [18, 19] proposed the so-called "hippocampal memory indexing theory" in which, to the contrary, the hippocampus proper plays a decisive role during retrieval. Teyler's theory assumes that for each complex event stored in the neocortex in a distributed manner a certain index (or: addressing mechanism) exists in the hippocampus that can be re-activated. By this, the hippocampus is able to point to the respective neocortical sites which, for their part, can quickly be re-activated to restore the whole memory pattern.

Finally, the work of the so-called Hippocampal Subfields Group has to be mentioned [20]. This is a sort of working group or league, currently consisting of 200 scientists from 18 countries. By using submillimetric high-field MRI the group has set itself the goal to standardize segmentation methods and newly define the anatomical boundaries of the MTL regions. A special focus is on the hippocampus proper and its

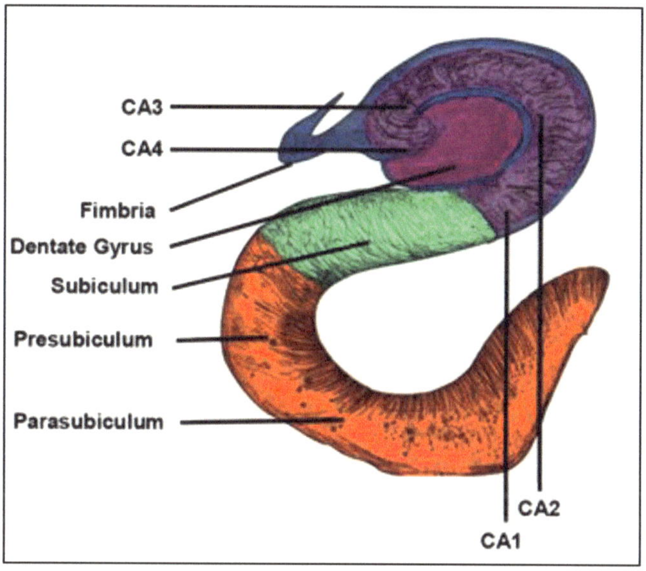

Figure 2.
Hippocampal subfields as displayed through a transverse section of the hippocampus. CA—cornu ammonis. (Kannappan et al. PLoS ONE 17(7), 2022).

subfields along the anterior-posterior axis, termed cornu ammonis (CA) fields 1, 2, 3, and 4, dentate gyrus (DG), parasubiculum, presubiculum, and subiculum (**Figure 2**). Interestingly, some new approaches are taken to build research on these subfields. For example, Cremona et al. [21] found a link between CA1 volume and the outcome of a learning test using word lists; in other words, subfield-level volumetry revealed a correlation between CA1 volume and the scores achieved by free recall, but not between the scores and the entire hippocampal volume.

2. The hippocampus in musical contexts

2.1 Studies on hippocampal plasticity

Neuroplasticity is one of the most widely studied issues in the field of "Music and Neuroscience." The term describes experience-driven reorganization in the human brain, mainly caused by hard exercise and extensive training (e.g., by repeating certain finger movements hundred times). Thus, the brain of practicing musicians is oftentimes regarded as a prototype model of neuroplasticity [22]. Changes in size and/or functional activity have mainly been studied on the macroscopic level, (but see [23] for a discussion of the underlying cellular and molecular mechanisms). For example in professional string players, sensorimotor areas are enlarged in the right half of the brain, because the trained left hand, that is, the fingering hand, has its cortical representations there [24]. Pianists, by contrast, reveal a larger anterior corpus callosum since piano playing requires precise coordination of the hands due to bimanual movements on the keyboard [25].

In a prominent MRI study on neuroplasticity, Maguire et al. [26] could show that the hippocampus is subjected to structural change too. However this time, plastic changes were caused by map-like knowledge and tremendous experience in navigation, that is, by spatial memory. What was examined there were the brains of a group of licensed London taxi drivers. In detail, the posterior parts of their right and left hippocampi revealed a significant increase in gray matter volume (in comparison with age-matched controls). (Note that activity in the taxi drivers' left hippocampus was interpreted as being caused by the innumerous social actions with which taxi driving is connected).

To further explore the processes of hippocampal reorganization, memory tasks in music are suitable too, both in regard to functional plasticity and structural plasticity. For example, Groussard et al. [27] studied semantic memory, that is, retrieval of musical facts such as style and/or composer. As stimuli, excerpts from the classical and modern repertoires were taken. Musicians and non-musician controls had to judge how familiar these excerpts were, using a four-point Likert-type scale for rating. (However, in order to study semantic memory more properly, an explicit naming task [e.g., which musical piece has just been heard] would be better.) Functional MRI revealed that musicians (in comparison with non-musician controls) had higher gray matter density in the left anterior hippocampus (also called hippocampal head) as well as increased functional activity in both the left and right hippocampi (**Figure 3**). Note that many studies testing semantic memory with authentic pre-existing musical stimuli have to deal (or struggle) with certain confounds. Because oftentimes, not only semantic content is remembered, but personal memories and emotional impressions are re-experienced too. In other words, semantic and episodic aspects cannot be properly disentangled. This, however, is less a methodological problem (or a design problem), but rather an inherent property of music itself and the effects it has on the listeners. Groussard et al. reported in regard to this point that the chosen "musical excerpts evoked personal memories in 85% of musicians, but only in 30% of non-musicians [27]".

A second point in neuroplasticity research is the time course, that is, the time frame in which structural reorganization occurs. To my knowledge, this issue was first addressed in a DC-EEG study conducted by [28]. In this study, Bangert and Altenmüller found co-activity in the auditory and sensorimotor areas already after 20 min of practice, that is, when beginners had had their first piano lesson.

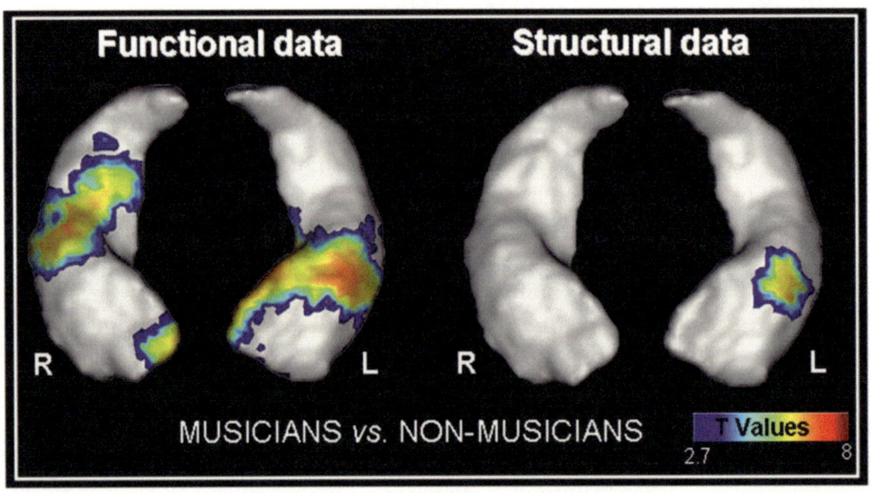

Figure 3.
When listening to familiar melodies both, semantic content and personal memories may come to mind. Plastic changes in the hippocampus are higher in musicians than in non-musicians. (Groussard et al. [27]).

Jäncke [29] writes the following: "Memory consolidation is time-dependent since the biochemical processes modulating synaptic processes need some time (at least 25 minutes) to develop and to install the new and altered synaptic contacts in the memory networks, including the release of various hormones into the bloodstream (i.e., epinephrine, norepinephrine, and cortisol) [29]."

Several other studies used diffusion tensor imaging (DTI) to study reorganization on the cellular level. This time, also hippocampal and parahippocampal structures were explicitly examined (e.g. [30]). Most interesting in this respect is a study by Jacobacci et al. [31]. It is a sort of extended replication using DTI and fMRI since the paradigm itself has been developed by Bönstrup et al. [16] before. In both studies, participants learned to tap a four-finger movement sequence (4-1-3-2-4), which had to be executed as quickly and precisely as possible. In short, beginners trained their motor skills. To see how dexterity improved on the micro-level the focus was on the average tapping speed within the test blocks and also on processes during the short rest periods in between; in other words, on improvements micro-online and micro-offline (see [16, 31] for further details). Interestingly, the hippocampus and precuneus were strongly activated during rest, that is, not during task execution, and this functional activity was immediately followed by plastic changes in the same anatomical regions. In contrast to that, cortico-cerebellar and cortico-striatal circuits were most active during task execution, that is, within the test blocks. So, two aspects are important here: First, rapid types of functional and structural reorganization begin about 20 or 30 min after the very first training, which confirms that consolidation is initially dependent on biochemical processes in the hippocampus [29]. Second, most crucial is the point that the hippocampus not only plays a decisive role in the consolidation of declarative memories but also in that of procedural content, that is, when motor sequences are learned and made more stable. In consequence, this enables musicians to recall finger movements automatically and in the correct serial order. Jacobacci et al. [31], therefore, suggest a common hippocampal-based mechanism for both the formation of declarative and non-declarative memories.

2.2 Recognition memory for melodies and hippocampal involvements

Every time when listening to a musical piece one has come across before, two processes can be distinguished from each other—a pure "feeling of familiarity" on the one hand and the "remembering" of the musical details on the other hand. So, recognition memory consists of two components. Interestingly, both recognition types depend on the hippocampus, which has to be seen as a part of a larger cortico-subcortical network this time. Thus, melody recognition is another example revealing how the hippocampus reacts in the context of music. In detail, the feeling of familiarity is defined as a vague impression of somehow knowing a musical piece, and for estimating how familiar it sounds a 7- or 10-point Likert-type rating scale can be useful. Remembering, by contrast, is often tested with an old/new auditory recognition task to prove how well the details of musical structure have been kept in mind. In these auditory recognition tasks, musical motifs, or longer excerpts, are played again, either in the original version or with slight variations, and these target stimuli have to be compared with the imprints stored in memory. (Note that for reasons of clarity, brain regions other than the hippocampus are not described in this section here.)

Interestingly, a similar differentiation has been made in regard to episodic memory, that is, when personal events are retrieved with certain vividness. In episodic memory, the term "recollection" is used to describe the process of recall (and re-experiencing) as many contextual details as possible. So, distinctions are made between the "feeling of familiarity" on the one hand and "recollection" on the other hand. There is a strong consensus that the latter, recollection, is hippocampal-dependent; however, in terms of the vague feeling that some episodic memories seem familiar it is not clear if this is hippocampal-dependent or rather based on activity in the perirhinal cortex (see, e.g., [32] for further details).

Back to the feeling of familiarity in music, Plailly et al. [33] found evidence for the hypothesis that all types of vague impressions have a common neural source, based on the idea that the feeling of familiarity is quite similar across all sensory modalities. They tested this common source hypothesis with fMRI and two modalities, music and odors. Participants had to judge whether musical excerpts (instrumental) and odors were familiar to them or not. An underlying common brain network could indeed be identified, whereby brain reactions to familiar sounding music were stronger than those to familiar smelling odors, for example those of herbs or flowers. Interestingly, a distinct left-hemispheric tendency could be observed that also referred to the left hippocampus and the left parahippocampus. So, for this bimodal type of stimulation a large overlap in (left-hemispheric) neural substrates has been found (including the hippocampal formation), which suggests that the feeling of familiarity is multimodal indeed, or, one could also say: modality-independent.

In contrast to that, Watanabe et al. [34] conducted an fMRI experiment to find out how well the details of musical structure could be remembered. So, in Ref. [34] "remembering," that is, the second hippocampal-dependent type of recognition memory, was examined, and old/new-decisions had to be made. To my knowledge, this fMRI study was the first in which co-activation of personal memories ("this reminds me of ...") was strictly minimized, since new musical stimuli had been generated with auto-composing software, by which any type of additional memory or association was kept at a minimum. During encoding, participants listened to 20 of these pieces (each of 3 s length, repeated three times). Then, 10 novel examples were added, and the subjects had to decide which stimulus was old (i.e., previously heard), and which was new. Brain activation was found in the right hippocampus for having correctly recognized pieces as old, so thus again retrieval success could be proven (cf. e.g., [3]).

Retrieval success was also the topic of a recent MEG/MRI study conducted by Bonetti and colleagues [35]. Again, old-new decisions had to be made; however, this time, the task was more demanding: During learning (or: encoding) participants listened four times to the right-hand part of a Bach Prelude (BWV 846) and also four times to its atonal counterpart. However, during retrieval, only short "snippets" (5-tone motifs) were presented. Half of these motifs were taken from the memorized prelude, whereas the other half was newly invented. Thus again, old examples had to be distinguished from novel ones, and this was done separately for the tonal and atonal versions. Overall, strong brain activity was observed in left MTL regions, including the left hippocampus and the left parahippocampus. Interestingly, this was only the case when the motifs were tonal and recognized as old, that is, as a part of the original prelude. Memorized atonal motifs, by contrast, evoked activity in right-hemispheric auditory regions, that is, without any significant functional change in the right hippocampus. In this study, retrieval was also examined in more detail in that event-related fields were analyzed in two frequency bands, a slow band (0.1–1 Hz) and a faster one (2–8 Hz). In the slow frequency band, voxel number was high for memorized tonal motifs from the third tone on, whereas for memorized atonal types voxel number was already high at motif beginning. These details suggest that in regard to tonal motifs, the hippocampus reacts to the whole, the entity or: Gestalt, in accordance with the observation that the hippocampus is engaged in the processing of patterns or "scenes" [cf. [7]]. Atonal motifs, by contrast, were processed tone-by-tone by activating large parts of the auditory network. Obviously, participants use different listening strategies during motif recognition: holistic whenever the replayed motifs are tonal (activating the hippocampus), and punctual whenever the motifs are atonal (activating the auditory network, including the insula).

Both studies on recognition memory [34, 35] reveal that the time windows between learning and recognition were very short. In other words, the test phase was immediately after encoding, with a delay of 15 min at the most (cf. [34]). This calls into question that the hippocampus is exclusively involved in long-term memory processes. Some researchers (e.g., [36, 37], also [5]) argue in favor of a more flexible approach. They understand hippocampal functioning beyond strictly set time frames through which by definition (hippocampal-independent) working memory and (hippocampal-dependent) long-term memory processes can be distinguished from each other. For example, Jeneson and Squire [36] point out (note that this is from a working memory perspective) that some task requirements; for example, greater memory load or maintenance of some complex information may occasionally exceed the capacity of working memory, and in such cases, the hippocampus is needed to support performance. Accordingly, in regard of working memory processes two new terms are suggested [36]: "subspan memory" and "supraspan memory"—the first for all those cases in which the storage capacity of short-term memory is sufficient. And the latter, supraspan memory, for the opposite, that is, when also long-term memory is needed to solve the task for which activity can be observed in the hippocampus.

3. Expert memory: strategies for memorizing and recall

3.1 Long-term working memory (LT-WM), chunking, and memory subtypes

Interestingly, this flexible interaction between working memory and long-term memory (plus hippocampal activity as just described) (see Section 2.2) is one out of three fundamental mechanisms by which the effectiveness of expert memory can be explained when skillful pieces have to be played by heart. So, in this section, I will elaborate on these three points to describe in detail the mechanisms underlying expert memory. These points are: (a) the use of a so-called retrieval scheme (which is based on flexible long-term/working memory interaction processes), (b) a principle called chunking as well as (c) different memory subsystems needed for expert performance.

a. Today, several studies (or: authentic reports) exist in which professional musicians describe their steps when preparing a new recital program. A successful example is the report series by Roger Chaffin (US-American psychologist) and Gabriela Imreh (US-pianist, of Romanian origin, e.g., [38], also [39]). According to them, a large part of daily practice consists in acquiring a so-called "retrieval scheme," which is part of conceptual (or: declarative) memory. For this, a musical piece is first partitioned into several segments (or: sound portions), and for each of these segments, a self-set landmark (i.e., a cue or an opener) is consciously memorized too. (Cues can be, for example, a structural detail, a sound, or a tricky fingering.) During recall, that is, during performance on stage, soloists just proceed from cue to cue, present in working memory, whereby each sound portion, stored in long-term memory, is directly accessed too. So, all in all, this mechanism makes fluent performance possible. For this linkage (or: interaction) between working memory and long-term memory, Ericsson and Kintsch coined the term "long-term working memory" (abbreviated "LT-WM" [1], see also [40]). However, to my knowledge, no study exists, in which LT-WM processes underlying time-precise sequence recall have been investigated with fMRI. At first glance, some striking parallels between LT-WM and Teyler's so-called "hippocampal memory indexing theory" might come to mind (see [18, 19]). According to the indexing theory, certain pointers are located in the hippocampus, through which the neocortical sites of a distributed array are re-activated so that the whole memory pattern can be restored. However, in regard to this parallel, some counterarguments exist: First, Teyler's theory is a neurophysiological one, based on the existence of reciprocal connections and synaptic physiology. LT-WM, by contrast, explains retrieval processes from a pure psychological point of view. A second point is that Teyler's theory provides the basis for episodic memory, enabling a subject to retrieve certain personal memories with all contextual details [19]. In other words: Teyler's theory might be suitable for a scene-like type of retrieval. However, to remember tone elements in serial order and precisely in time, as in performance situations, LT-WM seems to be the more suitable explanation: It describes a flexible, cue-triggered (WM-based) mechanism, making possible a rapid and reliable access to large amounts of domain-specific information, stored in long-term memory.

b. A second hallmark by which the efficiency of expert memory can be explained is a principle called "chunking," more precisely: a principle of thinking and performing in chunks. Chunks are defined as meaningful units of information (e.g., [41]), and the level of argumentation is working memory. In musical contexts, chunks often are pattern-like. Typical examples are short segments of musical scales that can be found in various musical pieces. From this follows that melodies are not processed tone-by-tone, but rather as groups or distinct entities, namely as chunks. Note that chunks can easily be detected in the written musical texts, the scores, since almost every composer adds certain slurs to indicate that a bunch of notes has to be played coherently as an entity, namely as a chunk or a musical phrase (see **Figure 4**). So, the key unit in music is the "musical phrase" rather than the chunk, also implying that every type of musical information is stored in musical phrases. Interestingly, it has been found that certain working memory processes occur at the boundaries of the musical phrases, and these transitional processes are more important than generally assumed. Because for grasping a melody as a whole, a first phrase has to be maintained in working memory, while the attention focus is directed to the next phrase. In other words, while playing or listening to melodies certain memory-and-attention-related micro processes occur, in particular at the edges of the phrases, the phrase boundaries. Interestingly, these transition processes in melodies do activate the hippocampus. Knösche et al. [42] found activity in the posterior parts of both hippocampi and also in parahippocampal areas, when one musical phrase (or chunk) was closed, and the next phrase was opened (for source localization EEG and MEG data were used) (see also [43]).

c. Finally, expert musicians rely on different memory types in parallel, by switching the attention back and forth between them. At least five subtypes can be distinguished from each other: Conceptual (or: declarative) memory in which the retrieval schemes and the details of music analyses are stored. Another subtype is auditory memory in which the entire musical piece is represented as

Figure 4.
A music example of high structural density, always played from memory during a recital: Frédéric Chopin Piano Sonata Nr 1, c minor, op. 4 (excerpt). Musical phrases, the chunks, are indicated by the larger slurs, whereas some of the smaller ones are signs for articulation. Some are also certain "ties," indicating that two notes of same pitch are attached.

a sound product. In addition, also motor, kinesthetic, and haptic subtypes exist, carrying memory details in terms of finger movements, muscle tension, and somatosensory impressions, for example, how it feels when strings are plucked. Some musicians also possess eidetic or photographic memory, which enables them to see the details of the score before the inner eye (see [38, 39] for further details). So, expert musicians memorize a musical piece in a multifaceted way. In other words, several backups exist in parallel, each with different pieces of information: auditory, conceptual, motor, haptic, kinesthetic, and/or visual. To my knowledge, studies, using fMRI and DTI, are still missing, in which the focus is set on how these memory subsystems interact with each other. However, in terms of music perception, that is, when subjects passively listen to music, some whole-brain connectivity studies do exist, and for this, all types of neural networks could be identified (e.g., see Alluri et al. [44]).

3.2 Mnemonic techniques: Method of loci

Whenever the to-be-remembered items have to be stabilized even further, people sometimes make use of a certain strategy of double encoding, commonly known as "mnemonic technique", for which hippocampal activity can be observed once again. Across the mnemonic types, the principle is as follows: (1) During encoding a primary item is tied (or: connected) to a secondary item, and this linkage is consciously memorized (e.g., [45]). Often, the primary item is complex or abstract; however, it is "the real thing," that is, the original piece of information that has to be remembered. The secondary item is a self-chosen adjunct, which, in most cases, is familiar to the subject and also pictorial in nature. (2) During retrieval, this additive, that is, the secondary item, can be easily remembered and by this, the primary item is activated too. In general, mnemonic techniques are used for maintaining sequence order, for example, when 100 words have to be recalled in the same order as they were presented. It becomes clear that connections are mostly one-to-one; that is, assignments are made between one primary and one imagined item [45]. Some types of mnemonic techniques exist in which both first and second items are merged, that is, integrated into a new unconventional entity (or: "composite" [46]), for which bizarre imagery is explicitly required [47]. It should also be mentioned that all types of mnemonic techniques are bi-modal; that is, both the primary item and the adjunct are taken from different domains (e.g., digits and pictures, faces and objects, words and places). The main point here is that these bi-modal connections (or: "between-domain associations" [48]) are produced not before reaching the hippocampus proper, that is, both items converge into the hippocampus, and bi-modal binding occurs exclusively there (see Mayes et al. [48] for further details). Obviously, this is not the case for "within-domain associations" (face-face, object-object, word-word), which "converge sufficiently in the perirhinal cortex to create a [...] memory representation [48]." Zeineh et al. [49] could demonstrate with fMRI that during bi-modal binding different hippocampal subfields are active, depending on whether data acquisition is during encoding or during retrieval. In this study [49], face-name associations were built and retrieved; in other words, names had to be assigned to new faces. During encoding, a complex of adjacent regions was active, in detail, CA fields 2 and 3 and the dentate gyrus (in short: CA2, CA3, DG), whereas during retrieval, changes in activation were primarily found in the posterior subiculum. (Note that face-name pairing tasks are quite frequent in everyday life situations, e.g., when a teacher has to memorize the names of all schoolchildren of a new class after summer vacation).

The oldest mnemonic strategy is the famous "method of loci," going back to Greek and Roman antiquity. For example, Marcus Tullius Cicero used this method for learning his long speeches by heart. The method of loci is based on visuospatial imagery and mental navigation, and the procedure is as follows: (1) During encoding, a familiar environment is chosen in mind (sometimes also physically), for example, rooms within a house or paths through the city or countryside. Anchoring (or: encoding) is in such a way that the primary items are put at distinct places along this path. (2) During retrieval, one simply imagines walking along the path, while, simultaneously, the anchored primary items come to mind in proper order. This method is useful when word lists and other types of sequential information have to be remembered. Cicero described the method of loci in detail in his *De oratore*, a textbook on rhetoric, which is one out of three sources informing about this "ars memoriae" in ancient times [50] (Note that the word mnemonic technique (or: mnemotechnic) was coined only in the nineteenth century.) It is said that Cicero used certain buildings of the Forum Romanum for this linkage which helped him remember the main points of his speeches reliably and in proper order when speaking to the audience [51] (**Figure** 5).

Even though the method of loci is mainly used for memorizing word lists or large amounts of text, musicians occasionally make use of it too (violinist, personal communication). While preparing for a concert she physically moved from corner to corner of her practicing room, each time 'depositing' a certain amount of sound information in one corner. During the concert, she simply imagined walking along the same path. Given that for this type of linkage a path through nature is chosen, an additional effect could be to feel mentally (and physically) relaxed. In other words, during recall, while performing a concert on stage, an imagined path through a natural environment could reduce inner tension. Since violinists play pieces from memory in upright position, this also raises the question whether the method of loci supports some egocentric (or: trunk-related) processes which, however, are located in the parietal lobe (for further details on egocentric frames and representations, see [11]). If this were the case, the retrosplenial cortex might play

Figure 5.
Cicero described the "method of loci" in De oratore, a textbook on rhetoric. For anchoring, he himself used certain buildings of the Forum Romanum which helped him remember the main points of his speeches in proper order.

a role in mediating between the hippocampus and the superior parietal lobe (SPL). The latter, SPL, is a multimodal spatial reference system into which visual, auditive, vestibular, tactile, kinesthetic—and, possibly, geocentric spatial—inputs converge for purposes of proprioception, that is, to bring room coordinates and one's own body position into balance (I do not want to elaborate on this further, since, to my knowledge, this lacks of empirical proof). What has been shown, however, is that after six weeks of daily mnemonic training with the method of loci (a web-based training platform) memory performance of naïve subjects increased significantly. More importantly, significant changes in network connectivity were found in the left hippocampus and bilateral retrosplenial cortex (cf. fMRI study by Dresler et al. [52]; see also [53] for similar results).

3.3 Mnemonic techniques: playing the North Indian tabla as a special case

In many non-Western cultures, especially India, mnemonic techniques are even more essential since in these countries, music does not exist in written form as scores, but rather it is transmitted aurally/orally [54]. Although mnemonic systems are quite similar across these non-Western cultures, the basic mechanism is entirely different from the principles on which the method of loci is based, namely spatial navigation and visual imagery. In non-Western cultures, connections are sound-to-syllable, that is, the sounds of a musical instrument are combined with certain syllables which are thought and/or spoken simultaneously while playing. At first glance, these syllables seem to be nonsense and arbitrary chosen. However, spectral analyses reveal that musical sounds and the syllables share common acoustic properties (see [55] for further details). Regarding non-Western traditions, the interesting point is the aspect of weighting or primacy, in other words, how sound-syllable connections should be understood. Which element is the predominant one—the musical sound or the speech syllable, that is, the first or the second item? This will be discussed in the following by taking as an example the North Indian tabla tradition. A tabla is a pair of hand drums of different size and pitch played in North India (**Figure 6**). Singers and sitar players are accompanied by this percussion instrument in small ensemble formations typical of the North Indian (Hindustani) classical music tradition.

There are many different ways of striking the tabla, however each time, a single drum sound is linked with a certain spoken syllable, called bol (from bolna, a Hindi verb for "to speak" [56]). Example syllables are: dha, dhin, dhun, tin, tun, kat, ghe, tra, kra [55, 56], thus, appearing as a sort of onomatopoeic equivalent for the drum sounds. However, the important aspect in terms of weighting or primacy is that bol syllables should not be considered as mere adjuncts. Because in India, the human voice is regarded as the origin of all music, the instrumental and the vocal forms, by which cosmic energy flows through the chest and the lungs to be formed by the vocal cords, the mouth, and the lips. Manuel and Blum [57] put it this way: "Musical sound [...] proceeds along a spiritual pathway from an unmanifested ideal form, through the navel, heart, throat, and finally the mouth. Vocal music, generated by vital breath [...], was conceived as a sublime manifestation of *nâda brahma*, a sort of primordial and divinely animated substratum of cosmic sound. [57]." Accordingly, vocal utterances are of particular value, and this holds also for the bol sequences. In this context, Rowell [58] writes the following (though the focus seems to be more on the South Indian counterpart, a syllable system called solkattu, [59]): "These syllables are more

Figure 6.
The tabla is the typical percussion instrument of North India: A pair of hand drums of different size and pitch is played by one musician seated on the ground. Different drum sounds (depending on where and how the drums are struck) are linked with spoken syllables, called bols, resulting in certain sound-to-syllable connections. Playing the tabla is a special case of using mnemonic techniques.

than drum syllables - they are abstract phonetic patterns that can be recited, played, and danced. [...] They are not medium-specific." So, the question about weighting or primacy can be answered in that way that drum sounds and phonetic elements should be considered as equivalent.

Regarding this sound-syllable system the wide range of applications becomes evident when having a look at the non-Western aural/oral teaching practice. This will be briefly explained by taking as an example the music education in Japan. Both Hughes [60] and Shehan [54] report that beginners learning to play the nōkan (a traditional Japanese flute) are taught to—first of all—sing the syllables in pure form before being allowed to pick up the flute and play the melody. (Note that in the Japanese nohkan tradition, the mnemonic syllables "hya" and "hyo" are central.) Obviously, the same holds for children in North India learning to play the tabla. Farrell [56] describes this learning process as "saying bols, writing them down, translating these into finger movements," meaning that the first step consists in building a mental representation of the sound syllables—enabling the novices to think the mnemonics fluently—before hand movements, that is, motor practice is added. Farrell continues: "The process of thinking, saying, and playing in tabla [sic] provides a fascinating example of how the building blocks of a music are conceptualized in thought and realized in sound [56]."

Another aspect worth considering is whether it is the *sound* of the tabla to which the spoken syllable is connected or the drum stroke itself (cf. [55]). In case of the latter, a close link would exist between the syllable and the sound-producing action. If so, the (chronological) order would be: "bol - action - sound" which is in line with the ancient Sanskrit wisdom that "the Word is translated into action [61]." From a neuroscience perspective, this might go into the direction of the human mirror neuron system: It is common knowledge that a special subset of audiovisual mirror neurons does not only respond when certain hand actions or gestures are observed, but also when the sounds resulting from these hand actions are heard [62, 63]. For listening to these action-related sounds, a left-hemispheric frontoparietal motor-related network has been identified [64], including Broca's area and the premotor cortex. This might also be active when bol sequences are played on the tabla, that is, when drum strokes (as sound-producing actions) are combined with syllables (the speech sounds). Note that with "tonic sol fa," or "solfège", a similar sound-to-syllable system exists in the Western culture. The principle is that an easily singable syllable, for example, do, re, mi, fa, or sol, is assigned to each of the seven pitches of the diatonic scale. However, in contrast to non-Western systems, these pitch-syllable pairs do not have any acoustic similarities. The practice of this solfège system can be traced back to Guido d'Arezzo, a famous music theoretician in the first half of the eleventh century, and it is mainly used for intonation exercises in singing lessons and for training auditory imagery until the present day (see, e.g., Hughes [60] for more details on Western arbitrary versus non-Western non-arbitrary sound-to-syllable systems).

4. What is memory for? Some reflections from the perspective of culture

So far, it has been shown that the hippocampus is one of the most relevant brain regions in humans, since almost every memory process—declarative and non-declarative—depends on hippocampal activity. Furthermore, to illustrate how sensitive the hippocampal region in terms of time-related micro-processes is, expert memory in music served as an example, manifesting itself in pitch- and time-accurate sequence recall. In this section, this data-driven approach is put aside in favor of reflections on memory from the viewpoints of anthropology and culture. What is memory for? What is its value? While searching for answers it makes sense to distinguish between the individual and the collective level. Regarding the first, the individual level, at least two points should be mentioned here: First, memories shape our character. That is, each remembered experience enriches the self, making each human unique, that is, different from others. Personal memories also guide people during decision making and help them avoid making the same mistake over and over again. Accordingly, memories play a decisive role to give persons a sense of identity. Second, semantic memory, that is, facts, and also know-how are some sort of preconditions for being creative. In other words, a fund of knowledge is beneficial for creative problem solving, enabling persons to build associations and have new ideas. For instance, during his first stay in Italy (1769–1771), the young (13-year-old) Mozart seized the opportunity to study counterpoint with Padre Martini in Bologna, a famous music theoretician at that time. Thus, although knowledge is often domain-general, the domain-specific type is even more useful, for example, to improve one's compositional style.

In the middle of the 1980s, human memory also became a topic in the socio-cultural sciences. Assmann [65] differentiates between memory as an "outer or: external phenomenon" (studied by cultural scientists) and memory as an "inner phenomenon" (studied by neuroscientists and psychologists). Moreover, den Boer [66] set both memory types, the internal and the external, in relation to each other. He compared the method of loci (the internal type, located in mind) with the so-called lieux de mémoire (the external type), that is, with places of memento in reality, such as certain monuments or holy places. (NB lieux de mémoire is a term coined by Pierre Nora, a French historian, in the late 1970s). In terms of inner and outer, den Boer [66] has the following opinion: "For the ancients, the *loci memoriae* were a necessary mnemotechnics in a society without modern media [...] For Cicero and Quintilian the *loci memoriae* were practical mental tools, free of ideology. [...] Nora's *lieux de mémoire* are also mnemotechnical devices, but [...] far from being neutral or free of value judgments. Most *lieux de mémoire* were created, invented, or reworked to serve the nation-state."

First attempts to consider memory as something social were made in the 1920s by Maurice Halbwachs, a French sociologist. He coined the term "mémoire collective" (collective memory), which should, more or less, be understood as the common knowledge of society. In other words, specific memories are of concern to a group or the entire society, and, to some degree, they are also present in the mind of each individual. Several measures are taken to keep these group and nation memories alive; for example, minutes of silence or days of remembrance to honor those who were killed in war or by assassinations, but also national and religious holidays belong to this category. In short, the past is tied to the present; that is, collective memory is refreshed by repetition at regular intervals, mainly from year to year. Thus, joining festivities or commemorative ceremonies may strengthen a person's cohesion with the group, in other words: his or her cultural and/or national identity.

Another important point is that, since ancient times and throughout all societies and ethnic groups, certain persons stand out from the crowd through calling myths, heroic epics, and/or group-related events back to mind by transmitting them orally, for instance, by telling these stories on market places. The aim is to strengthen the audience's cultural identity. Assmann [65] explains this point by taking as an example the "griots," a caste of wandering singers in Africa. But also shamans, bards, heads of tribes, and, in a broader sense, priests and teachers are good examples. All have in common to be explicit carriers of internalized knowledge (somehow serving as a medium), and with this, they help preserve oral traditions and culture. What they communicate are, for the most part, words of wisdom and insights, that is, facts about the past, what had happened to a tribe, combined with own experiences and some advice for the future. Walter J. Ong, an American media theorist [67], set out that griots, bards, and other carriers of wisdom make use of certain poetic devices (functioning as a sort of mnemonic device) to transmit their narratives properly from generation to generation. Such poetic devices are, for example, rhymes, alliterations, assonances, formulaic expressions as well as many repetitions and redundancies regarding content. In consequence, a story can be clearly remembered and exactly repeated by the storyteller, while, at the same time, this is catchy for the audience. The effectiveness of such poetic devices was verified in a study by Tillmann and Dowling [68]. They conducted an experiment on short-term memory, using two kinds of verbal material, prose and poetry. In this study, short-term memory for surface details declined significantly later in poems than in prose. This is because in poetry, the metrical structure and the rhyming at the end of the verse lines tie phrases together, resulting in a coherent whole. So, in this study subtle distinctions could therefore be easily detected in poems, for example, when after delays of either 4 s or 30 s certain lures were presented that had to be distinguished from the original verse lines (verbatim) (see [68] for further details).

Regarding oral traditions, a further point is of utmost importance: What is passed on by word of mouth? Is it the gist, that is, the storyline, in other words the central plots and aspects leaving space for improvisation? Or is it that all the details can be precisely remembered and recited even three or four generations later? The answer goes in both directions, since this depends on the type of narrative and on the efforts made during encoding. Anyhow, when holy wisdom has to be transmitted in religious contexts oftentimes the latter is the case. Rowell [61] describes the process of verbatim recall as follows (referring to collections of hymns from the *Rgveda*): "Young trainees were put through memory-building routines that boggle the mind: recitation of the text both with and without consciousness of its meaning, recitation both forward and backward, recitation both with phonetic junctures [...] between words and with separations between words, metrical recitation and recitation in a continuous flow (except for the obligatory caesuras), and recitation of pairs of syllables in distorted [...] sequences. [...] The aim was to instill an automatic and total command of the text that would rule out even the slightest possibility of error." Assmann puts it this way [65]: "The Vedas are not written down because Brahmans trust in written texts less than in memory. Each holy text is a kind of spoken temple, making the Holy present through the medium of voice."

From the neuroscience point of view, an interesting result was obtained in a study by Robin and Moscovitch [69]. There, hippocampal activity could be observed for both—the gist and the details, in other words, for coarse-grained and fine-grained aspects. An example is: "party" (gist) vs. "cake at 10th birthday party" (details of a single episode). In terms of this, the hippocampus shows activity in different subregions along the long-axis. Robin and Moscovitch [69] propose the following: "Perceptually detailed, highly specific representations are mediated by the posterior hippocampus and neocortex, gist-like representations by the anterior hippocampus [...] These representations can co-exist and the degree to which each is utilized is determined by its availability and by task demands." Obviously, this seems to be valid for both processes, encoding and retrieval, showing once again, the dynamics of the hippocampus proper.

Another example of verbatim recall can be found in the religious practices of the Islam: The Quran is considered as the word of God, that is, as literal truth, and any type of misquoting, or passing it on in incorrect form is considered a sin [70]. Furthermore, the Quran is recited in in its original language, Classical Arabic, without exception. In other words: no translations are allowed which is in stark contrast to Christianity. Since best effects are achieved in aural/oral form, Muslims prefer reciting (and memorizing) to reading the Quran in silence. Thus, religious Muslims make learning the Quran their life's work, starting at the age of four or five when attending the Islamic preschools and the Quranic schools [71]. This exemplifies that every religious Muslim is the carrier of (internalized) Quranic knowledge, guiding him (or her) through life as a sort of moral compass. Boyle [71] speaks of "Quranic memorization as a process of embodiment." In this case, however, learning is mostly by rote, that is, listening to the Quran teachers (and/or reading the Quran), and then repeating and rehearsing the text word by word and line by line ([71], see **Figure 7**). Some Muslims also make use of their eidetic or photographic memory through encoding the Arabic calligraphy too. Saleem [70] interprets this as a mnemonic mechanism by which the pictorial calligraphic characters are the secondary items so that each sura is memorized in the form of sound-picture combinations or "letter-sound relationships [70]," to use his own words.

However, the interesting point is here that Quran memorization is purely through the sound of the language, and for this, the phonetic characteristics of Classical Arabic are crucial. That is, no semantic analysis or deeper understanding of the meaning is required [70–72]. This special, sound-based practice of memorizing has to be seen against the background of religious aesthetics. Navid Kermani, a cultural researcher of Persian origin, puts emphasis on the point that the Quran should be experienced in an aesthetic (and acoustic) manner, that is, sensually, in a poetic way through the attractiveness of sound. He has elaborated on this in

Figure 7.
Memorizing the suras of the entire Quran is a lifelong task, starting at the age of four or five when attending Islamic preschools and the Quranic schools. In principle, every religious Muslim is the carrier of (internalized) Quranic knowledge, guiding him (or her) through life as a sort of moral compass.

his book *God is beautiful. The aesthetic experience of the Quran* ([73], see also [74]). Interestingly, a similar tendency prevailed in the Christian Occident during the Middle Ages. Yates [50] reported that in the time of Albertus Magnus and Thomas Aquinus (both thirteenth century) the mnemonic techniques moved over from rhetoric to ethics and have been regarded as part of virtue theory ever since.

5. Conclusions

In this book chapter, playing music by heart served as an example to illustrate that sequence recall is dependent on the hippocampus, a brain region within the medial temporal lobe. In detail, hippocampal activity was evoked by micro-processes at phrase boundaries [42], during motor sequence learning [31], and when detecting irregular offset-onset intervals [4]. Thus, emphasis was on expert memory, that is, on entirely different aspects than those in episodic memory, when certain unique experiences are remembered (e.g., [5, 6]). This performance example has also shown that hippocampal activity is needed every time when memory processes are required in the performing arts, that is, in contexts specific to humans. On the other hand, numerous cross-species studies exist, in which the hippocampi of birds, rodents, and monkeys have been studied in terms of anatomy, functioning, and connectivity, illustrating that memory is also of immense value for survival [cf. [11]]. For instance, sparrows or squirrels, remembering during winter-time where they have laid in their supply of nuts, are good examples to show how vividly important spatial memory is from a Darwinian point of view.

In this overview chapter, it has also been shown how dynamic and flexible the hippocampal brain region is, oftentimes, because different hippocampal subfields along the long-axis are active. Example processes, illustrating this flexibility, are: (1) changes between the gist and the details [68, 69], (2) flexible interactions between long-term and working memory [1, 36], (3) reactions to retrieval success and to the opposite, that is, novelty detection [4, 34, 35] as well as (4) reactions to the feeling of familiarity and the remembering of structural details [33–35]. In many studies, hippocampal reactions were also compared between groups, in particular between musicians and non-musician controls, showing that the hippocampus of musicians is not only a prototype example in regard to neuroplasticity but also in terms of many other respects. (Still, hippocampal plasticity is one of the most studied topics in the field of Music and Neuroscience (e.g., [27]).) The disadvantage is that, until now, there are many gaps in research; that is, functional MRI data are missing on how the hippocampus might react to musical aspects other than time, making it unavoidable to sometimes argue by analogy. In other words: many suggestions need verification.

Occasionally, musicians make use of the (hippocampal-based) method of loci, the oldest mnemonic strategy based on visuospatial imagery and mental navigation. For all mnemonic techniques, bi-modality is characteristic, meaning that both the primary item and the adjunct are taken from different domains to stabilize content by way of double encoding. Zeineh et al. [49] could demonstrate with fMRI that during bi-modal binding different hippocampal subfields are active, depending on whether encoding or retrieval processes are examined. Further research on hippocampal sub-fields (see, e.g., [21]) is a promising avenue to explore the functioning of the hippo-campus in all its detail. Mnemonic techniques are even more essential in non-Western cultures, since there music is frequently transmitted in aural/oral form [54]. This has been exemplified by the North Indian tabla tradition. Whether and to what extent the hippocampus is involved when drum patterns are played and certain onomatopoeic syllables are thought or spoken simultaneously is a matter for future research, and suitable paradigms have to be developed to disentangle the process of drum sequence playing from that of syllable speaking.

Further points of interest are implications regarding culture, more precisely: which relationships exist between cognitive neuroscience of memory on the one hand and the socio-cultural sciences on the other hand. In this respect, the following points are important: (1) memories shape the self and make humans unique; (2) knowledge stored in semantic memory is beneficial for creative problem-solving; (3) collective memories strengthen a person's cohesion with the group, that is, his or her cultural and/or national identity, (4) throughout all ethnic groups shamans, bards, priests, and teachers are explicit carriers of internalized knowledge, thus helping to preserve oral traditions and culture; (5) religious knowledge, when internalized, may serve as a moral compass for each individual, this is especially true for religious Muslims memorizing the Quran their whole life.

Thus, one can conclude that the hippocampus is indeed one of the most relevant brain regions in man, having effects on almost every aspect of human life from basic survival to moral behavior and the preservation of culture.

Conflict of interest

None. I am the single author.

Author details

Christiane Neuhaus
Institute of Systematic Musicology, University of Hamburg, Hamburg, Germany

*Address all correspondence to: christiane.neuhaus@uni-hamburg.de

References

[1] Ericsson KA, Kintsch W. Long-term working memory. Psychological Review. 1995;**2**(2):211-245. DOI: 10.1037/0033-295X.102.2.211

[2] Barnett AJ, O'Neil EB, Watson HC, Lee ACH. The human hippocampus is sensitive to the durations of events and intervals within a sequence. Neuropsychologia. 2014;**64**:1-12. DOI: 10.1016/j.neuropsychologia.2014.09.011

[3] Meltzer JA, Constable RT. Activation of human hippocampal formation reflects success in both encoding and cued recall of paired associates. NeuroImage. 2005;**24**:384-397. DOI: 10.1016/j.neuroimage.2004.09.001

[4] Herdener M, Esposito F, di Salle F, Boller C, et al. Musical training induces functional plasticity in human hippocampus. The Journal of Neuroscience. 2010;**30**(4):1377-1384. DOI: 10.1523/JNEUROSCI.4513-09.2010

[5] Olsen RK, Moses SN, Riggs L, Ryan JD. The hippocampus supports multiple cognitive processes through relational binding and comparison. Frontiers in Human Neuroscience. 2012;**6**:1-13. DOI: 10.3389/fnhum.2012.00146

[6] Backus AR, Bosch SE, Ekman M, Grabovetsky AV, Doeller CF. Mnemonic convergence in the human hippocampus. Nature Communications. 2016;**7**:11991. DOI: 10.1038/ncomms11991

[7] Maguire EA, Mullally SL. The hippocampus: A manifesto for change. Journal of Experimental Psychology: General. 2013;**142**(4):1180-1189. DOI: 10.1037/a0033650

[8] Maguire EA, Valentine ER, Wilding JM, Kapur N. Routes to remembering: The brains behind superior memory. Nature Neuroscience. 2003;**6**(1):90-95. DOI: 10.1038/nn988

[9] Lavenex P, Amaral DG. Hippocampal-neocortical interaction: A hierarchy of associativity. Hippocampus. 2000;**10**:420-430. DOI: 10.1002/1098-1063(2000)10:4<420::AID-HIPO8>3.0.CO;2-5

[10] Damasio AR. The brain binds entities and events by multiregional activation from convergence zones. Neural Computation. 1989;**1**(1):123-132. DOI: 10.1162/neco.1989.1.1.123

[11] Burgess N, Maguire EA, O'Keefe J. The human hippocampus and spatial and episodic memory. Neuron. 2002;**35**:625-641. DOI: 10.1016/S0896-6273(02)00830-9

[12] Strange BA, Witter MP, Lein ES, Moser EI. Functional organization of the hippocampal longitudinal axis. Nature Reviews Neurscience. 2014;**15**:655-669. DOI: 10.1038/nrn3785

[13] Persson J, Söderlund H. Hippocampal hemispheric and long-axis differentiation of stimulus content during episodic memory encoding and retrieval: An activation likelihood estimation meta-analysis. Hippocampus. 2015;**25**:1614-1631. DOI: 10.1002/hipo.22482

[14] Robinson JL, Salibi N, Deshpande G. Functional connectivity of the left and right hippocampi: Evidence for functional lateralization along the long-axis using meta-analytic approaches and ultra-high field functional neuroimaging. NeuroImage. 2016;**135**:64-78. DOI: 10.1016/j.neuroimage.2016.04.022

[15] Ritchey M, Libby LA, Ranganath C. Cortico-hippocampal systems involved

in memory and cognition: The PMAT framework. Progress in Brain Research. 2015;**219**:45-64. DOI: 10.1016/bs.pbr.2015.04.001

[16] Bönstrup M, Itturate I, Thompson R, Cruciani G, Censor N, Cohen LG. A rapid form of offline consolidation in skill learning. Current Biology. 2019;**29**:1346-1351. DOI: 10.1016/j.cub.2019.02.049

[17] Takashima A, Nieuwenhuis ILC, Jensen O, Talamini LM, Rijpkema M, Fernández G. Shift from hippocampal to neocortical centered retrieval network with consolidation. The Journal of Neuroscience. 2009;**29**(32):10087-10093. DOI: 10.1523/JNEUROSCI.0799-09.2009

[18] Teyler TJ, DiScenna P. The hippocampal memory indexing theory. Behavioral Neuroscience. 1986;**100**(2):147-154. DOI: 10.1037/0735-7044.100.2.147

[19] Teyler TJ, Rudy JW. The hippocampal indexing theory and episodic memory: Updating the index. Hippocampus. 2007;**17**:1158-1169. DOI: 10.1002/hipo.20350

[20] Olsen RK, Carr VA, Daugherty AM, La Joie R, et al. Progress update from the hippocampal subfields group. Alzheimer's & Dementia: Diagnosis, Assessment & Disease Monitoring. 2019;**11**:439-449. DOI: 10.1016/j.dadm.2019.04.001

[21] Cremona S, Zago L, Mellet E, Petit L, et al. Novel characterization of the relationship between verbal list-learning outcomes and hippocampal subfields in healthy adults. Human Brain Mapping. 2021;**42**:5264-5277. DOI: 10.1002/hbm.25614

[22] Münte TF, Altenmüller E, Jäncke L. The musician's brain as a model of neuroplasticity. Nature Reviews Neuroscience. 2002;**3**:473-478. DOI: 10.1038/nrn843

[23] Zatorre RJ, Fields RD, Johansen-Berg H. Plasticity in gray and white: Neuroimaging changes in brain structure during learning. Nature Neuroscience. 2012;**15**(4):528-536. DOI: 10.1038/nn.3045

[24] Elbert T, Pantev C, Wienbruch C, Rockstroh B, Taub E. Increased cortical representation of the fingers of the left hand in string players. Science. 1995;**270**(5234):305-307. DOI: 10.1126/science.270.5234.305

[25] Schlaug G, Jäncke L, Huang Y, Staiger JF, Steinmetz H. Increased corpus callosum size in musicians. Neuropsychologia. 1995;**33**:1047-1055. DOI: 10.1016/0028-3932(95)00045-5

[26] Maguire EA, Gadian DG, Johnsrude IS, Good CD, Ashburner J, Frackowiak RSJ, et al. Navigation-related structural change in the hippocampi of taxi drivers. Proceedings of the National Academy of Sciences. 2000;**97**(8):4398-4403. DOI: 10.1073/pnas.070039597

[27] Groussard M, La Joie R, Rauchs G, Landeau B, et al. When music and long-term memory interact: Effects of musical expertise on functional and structural plasticity in the hippocampus. PLoS One. 2010;**5**(10):e13225. DOI: 10.1371/journal.pone.0013225

[28] Bangert M, Altenmüller EO. Mapping perception to action in piano practice: A longitudinal DC-EEG study. BMC Neuroscience. 2003;**4**(26):1-14. DOI: 10.1186/1471-2202-4-26

[29] Jäncke L. Music and memory. In: Thaut MH, Hodges DA, editors. The Oxford Handbook of Music and Memory. Oxford: Oxford University Press; 2019. pp. 237-262 ch11

[30] Sagi Y, Tavor I, Hofstetter S, Tzur-Moryosef S, Blumenfeld-Katzir T, Assaf Y. Learning in the fast lane: New

insights into neuroplasticity. Neuron. 2012;**73**:1195-1203. DOI: 10.1016/j.neuron.2012.01.025

[31] Jacobacci F, Armony JL, Yeffal A, Lerner G, et al. Rapid hippocampal plasticity supports motor sequence learning. Proceedings of the National Academy of Sciences. 2020;**117**(38):23898-23903. DOI: 10.1073/pnas.2009576117

[32] Squire LR, Wixted JT, Clark RE. Recognition memory and the medial temporal lobe: A new perspective. Nature Reviews Neuroscience. 2007;**8**(11):872-883. DOI: 10.1038/nrn2154

[33] Plailly J, Tillmann B, Royet J-P. The feeling of familiarity of music and odors: The same neural signature? Cerebral Cortex. 2007;**17**:2650-2658. DOI: 10.1093/cercor/bhl173

[34] Watanabe T, Yagishita S, Kikyo H. Memory of music: Roles of right hippocampus and left inferior frontal gyrus. NeuroImage. 2008;**39**:483-491. DOI: 10.1016/j.neuroimage.2007.08.024

[35] Fernández-RubioG,BratticoE,KotzSA, Kringelbach ML, Vuust P, Bonetti L. Magnetoencephalography recordings reveal the spatiotemporal dynamics of recognition memory for complex versus simple auditory sequences. Communications Biology. 2022;**5**:1-12. DOI: 10.1038/s42003-022-04217-8

[36] Jeneson A, Squire LR. Working memory, long-term memory, and medial temporal lobe function. Learning & Memory. 2012;**19**:15-25. DOI: 10.1101/lm.024018.111

[37] Yonelinas AP. The hippocampus supports high-resolution binding in the service of perception, working memory and long-term memory. Behavioral Brain Research. 2013;**254**:34-44. DOI: 10.1016/j.bbr.2013.05.030

[38] Chaffin R, Imreh G. Practicing perfection: Piano performance as expert memory. Psychological Science. 2002;**13**(4):342-349. DOI: 10.1111/j.0956-7976.2002.00462.x

[39] Fonte V, Pipa L, Williamon A, Lisboa T. Memorising contemporary piano music as described by professional pianists. Music & Science. 2022;**5**:1-15. DOI: 10.1177/20592043221132932

[40] Lehmann AC, Chaffin R. Erinnern und Wiedererkennen: Auswendig- und Vomblattspiel [Remembering and Recognition: Playing by memory and playing by sight]. In: Bruhn H, Kopiez R, Lehmann AC, editors. Musikpsychologie – Das neue Handbuch. Reinbek bei Hamburg: rororo; 2008. pp. 354-373 ch.4.4b

[41] Lehmann AC, Kopiez R. Sight-reading. In: Hallam S, Cross I, Thaut M, editors. The Oxford Handbook of Music Psychology. Oxford: Oxford University Press; 2009. pp. 344-351 ch32

[42] Knösche TR, Neuhaus C, Haueisen J, Alter K, Maess B, Witte OW, et al. Perception of phrase structure in music. Human Brain Mapping. 2005;**24**:259-273. DOI: 10.1002/hbm.20088

[43] Neuhaus C, Knösche TR, Friederici AD. Effects of musical expertise and boundary markers on phrase perception in music. Journal of Cognitive Neuroscience. 2006;**18**(3):472-493. DOI: 10.1162/jocn.2006.18.3.472

[44] Alluri V, Toiviainen P, Burunat I, Kliuchko M, Vuust P, Brattico E. Connectivity patterns during music listening: Evidence for action-based processing in musicians. Human Brain Mapping. 2017;**38**:2955-2970. DOI: 10.1002/hbm.23565

[45] Bower GH. Analysis of a mnemonic device. Modern psychology uncovers

the powerful components of an ancient system for improving memory. American Scientist. 1970;**58**(5):496-510

[46] Bellezza FS. Mnemonic devices: Classification, characteristics, and criteria. Review of Educational Research. 1981;**51**(2):247-275. DOI: 10.3102/00346543051002247

[47] Wollen KA, Margres MG. Bizarreness and the imagery multiprocess model. In: McDaniel MA, Pressley M, editors. Imagery and related mnemonic processes. Theories, individual differences, and applications. New York: Springer; 1987. pp. 103-127 ch5

[48] Mayes A, Montaldi D, Migo E. Associative memory and the medial temporal lobes. Trends in Cognitive Sciences. 2007;**11**(3):126-135. DOI: 10.1016/j.tics.2006.12.003

[49] Zeineh MM, Engel SA, Thompson PM, Bookheimer SY. Dynamics of the hippocampus during encoding and retrieval of face-name pairs. Science. 2003;**299**:577-580. DOI: 10.1126/science.1077775

[50] Yates FA. The art of memory. London: Routledge & Kegan Paul; 1966. p. 400

[51] Becchetti A. Hippocampal formation and the classical art of memory. Proceedings of the National Academy of Sciences. 2010;**107**(25):104. DOI: 10.1073/pnas.1004950107

[52] Dresler M, Shirer WR, Konrad BN, Müller NCJ, et al. Mnemonic training reshapes brain networks to support superior memory. Neuron. 2017;**93**:1227-1235. DOI: 10.1016/j.neuron.2017.02.003

[53] Kondo Y, Suzuki M, Mugikura S, Abe N, Takahashi S, Iijima T, et al. Changes in brain activation associated with use of a memory strategy: A functional MRI study. NeuroImage. 2005;**24**:1154-1163. DOI: 10.1016/j.neuroimage.2004.10.033

[54] Shehan P. The oral transmission of music in selected Asian cultures. Bulletin of the Council for Research in Music Education. 1987;**92**:1-14. Available from: www.jstor.org/stable/40318119

[55] Patel AD, Iversen JR. Acoustic and perceptual comparison of speech and drum sounds in the North Indian tabla tradition: An empirical study of sound symbolism. In: Solé MJ, Recasens i Vives D, Romero J, editors. Proceedings of the 15th International Congress of Phonetic Sciences. Barcelona; 3-9 August 2003. pp. 925-928

[56] Farrell G. Thinking, saying, playing: Children learning the tabla. Bulletin of the Council for Research in Music Education. 1997;**133**:14-19. Available from: www.jstor.org/stable/40318833

[57] Manuel P, Blum S. Classical aesthetic traditions of India, China, and the Middle East. In: Gracyk T, Kania A, editors. The Routledge Companion to Philosophy and Music. New York: Routledge; 2011. pp. 245-256 ch23

[58] Rowell L. The musical imagery of India. In: Godøy RI, Jørgensen H, editors. Musical Imagery. Lisse: Swets & Zeitlinger; 2001. pp. 289-302 ch16

[59] Dineen DF. Speaking time, being time: Solkaṭṭu in South Indian performing arts [PhD thesis]. Middleton, Connecticut: Wesleyan University. 2015.

[60] Hughes DW. No nonsense: The logic and power of acoustic-iconic mnemonic systems. British Journal of Ethnomusicology. 2000;**9**(2):93-120. DOI: 10.1080/09681220008567302

[61] Rowell L. Music and musical thought in early India. Chicago: University of Chicago Press; 1992. p. 409

[62] D'Ausilio A. Mirror-like mechanisms and music. The Scientific World Journal. 2009;**9**:1415-1422. DOI: 10.1100/tsw.2009.160

[63] Kohler E, Keysers C, Umiltà MA, Fogassi L, Gallese V, Rizzolatti G. Hearing sounds, understanding actions: Action representation in mirror neurons. Science. 2002;**297**:846-848. DOI: 10.1126/science.1070311

[64] Lahav A, Saltzman E, Schlaug G. Action representation of sound: Audiomotor recognition network while listening to newly acquires actions. The Journal of Neuroscience. 2007;**27**(2):308-314. DOI: 10.1523/jneurosci.4822-06.2007

[65] Assmann J. Das kulturelle Gedächtnis - Schrift, Erinnerung und politische Identität in den frühen Hochkulturen. [Cultural memory and early civilization: Writing, remembrance, and political imagination]. 7th ed. München: C.H. Beck; 2013. p. 344

[66] den Boer P. Loci memoriae – Lieux de mémoire. In: Erll A, Nünning A, editors. Media and Cultural Memory. Berlin: Walter de Gruyter; 2008. pp. 19-25 ch1

[67] Ong WJ. Oralität und Literalität - Die Technologisierung des Wortes. [Orality and literacy - The technologizing of the word]. 2nd ed. Wiesbaden: Springer Fachmedien; 2016. p. 190. DOI: 10.1007/978-3-658-10972-1

[68] Tillmann B, Dowling WJ. Memory decreases for prose, but not for poetry. Memory & Cognition. 2007;**35**(4):628-639. DOI: 10.3758/BF03193301

[69] Robin J, Moscovitch M. Details, gist and schema: Hippocampal-neocortical interactions underlying recent and remote episodic and spatial memory. Current Opinion in Behavioral Sciences. 2017;**17**:114-123. DOI: 10.1016/j.cobeha.2017.07.016

[70] Saleem A. Memorization without comprehension: A window onto the 'extremities' of the capability of human brain. Al-Idah. 2018;**36**(1):25-40

[71] Boyle HN. Memorization and learning in Islamic schools. Comparative Education Review. 2006;**50**(3):478-495. DOI: 10.1086/504819

[72] Eickelman DF. The art of memory: Islamic education and its social reproduction. Comparative Studies in Society and History. 1978;**20**(4):485-516. DOI: 10.1017/S0010417500012536

[73] Kermani N, Gott ist schön. Das ästhetische Erleben des Koran. [God is beautiful. The aesthetic experience of the Quran]. 5th ed. München: C.H. Beck; 2017. p. 546

[74] Brunner R. Review: Reviewed Work: Gott ist schön. Das ästhetische Erleben des Koran by Navid Kermani. [God is beautiful. The aesthetic experience of the Quran by Navid Kermani.]. Bulletin of the School of Oriental and African Studies, University of London. 2002;**65**(1):155-158. Available from: www.jstor.org/stable/4145912

Section 4

Hippocampal Plasticity

Chapter 6

The Modulation of Hippocampus Plasticity

Livia Livinț Popa, Diana Chira, Aida Asaftei, Stefan Strilciuc and Dafin Mureșanu

Abstract

The hippocampus is a brain region that plays a vital role not only in learning and memory but also in a variety of cognitive processes. Additionally, the hippocampus is known for its plasticity or its ability to adapt structural and functional properties in response to internal and external factors. This plasticity is intricately modulated by a variety of factors, including neurotransmitters (such as glutamate), neurotrophic factors (such as BDNF, IGF-1, VEGFα, and NGF), cytokines, chemokines, adipokines (such as leptin and adiponectin), and hormones (such as cortisol, beta-endorphins, thyroid hormones, and noradrenaline). Changes in the number, length, type, and shape of dendritic spines within the hippocampus can influence neurotransmission, and subsequently behavior, through modulation of glutamatergic neurons. There are several interventions, including pharmacological treatments (such as antidepressants or multimodal drugs) and non-pharmacological interventions (such as non-invasive brain stimulation of targeted regions, physical exercise, and an enriched environment) that promote neurogenesis in the dentate gyrus, resulting in beneficial effects on cognition and mood. Both types of therapies have the potential to increase connectivity between the hippocampus and other areas of the brain involved in motor and cognitive control, and thus, improve performance in specific tasks.

Keywords: hippocampus, plasticity, cognitive control, neurotrophic factors, non-pharmacological interventions

1. Introduction

The hippocampus is a crucial brain region that plays a vital role in learning and memory, particularly in the acquisition of new memories. It is also involved in the regulation of emotion and mood. The hippocampus is affected by aging and age-related cognitive disorders, such as the progressive atrophy of its volume.

The hippocampus is composed of a heterogeneous population of neurons that are distinguished by their age, morphological characteristics, and connectivity. It is anatomically organized into three main areas: CA1, CA3 (short for Cornu Ammonis), and the dentate gyrus (DG). Area CA1 is responsible for encoding memories while area CA3 is thought to mediate the retrieval of complete memories from partial information (pattern completion). The DG is important for spatial pattern separation, the process by which similar incoming information or stimuli are transformed into distinct, non-overlapping experiences. The hippocampus is thought to support the recollection of episodic events by representing the complex spatiotemporal patterns that uniquely define typical real-world events. Patients with lesions to the hippocampus have marked deficits in episodic and spatial memory [1–5].

This chapter focuses first on the modulation of hippocampal plasticity in terms of neurogenesis, structural, and functional plasticity, as well as on the relevant factors. It then examines how the connectivity between the hippocampus and other cerebral structures leads to the creation of new associations starting from familiar concepts or the construction of new concepts and categories. The second half of the chapter outlines the medical interventions designed to modulate hippocampal plasticity, including both pharmacological (such as therapy with Cerebrolysin, antidepressant medication, and even drug abuse) and non-pharmacological interventions (such as transcranial electrical stimulation, transcranial magnetic stimulation, and physical exercise).

2. Plasticity

2.1 Introducing neuroplasticity

2.1.1 From neurogenesis to reorganized or new connections

Neuroplasticity refers to the brain's ability to change and adapt with input and repetitive use. This capacity allows the brain to adjust to environmental changes, learn, and repair itself after lesions or disease. Genetic, neuronal, and neurochemical factors all play a role in neuroplasticity, and its limits can be manipulated through clinical and pharmacological interventions [1, 2]. The hippocampus is a highly adaptable structure that is particularly prone to neuroplasticity. This process involves neurogenesis, which is the proliferation and differentiation of neural stem cells (NSCs) and the subsequent changes in the morphology and activity of the differentiated neurons. In adult mammals, the subventricular and subgranular zones of the hippocampus are the two neurogenic niches where NSCs proliferate and differentiate to generate new neurons [2, 4].

The central nervous system undergoes structural and functional changes during development years and adulthood as it is constantly stimulated by internal physiology and environmental conditions. The hippocampus is one of the few areas in the brain where neurogenesis persists throughout adulthood, supporting learning and memory and potentially contributing to brain repair [1, 4]. While the regenerative potential of stem cell niches in the brain is still a topic of debate, an increasing body of evidence suggests that in the hippocampus, newborn neurons integrate into existing circuits and play a critical role in learning, memory, and neurological disorders [4].

Understanding hippocampal neurogenesis is therefore of utmost importance, due to its crucial role in cognition and regulation of emotions, as new neurons are integrated rapidly into its network, providing a certain level of cellular flexibility to the brain [3].

2.1.2 Types of neuroplasticity

Neuroplasticity, the ability of the brain to change and adapt with input and repetitive use, is an extraordinary tool that allows the mature brain to adjust to environmental changes, to learn, and to repair itself after lesions or disease. This capacity is divided into two categories: structural and functional plasticity.

Structural plasticity involves the expansion or retraction of the synaptic area through the remodeling of spines, dendrites, and/or axons. It involves dendritic formation and spine development, which are regulated through various cellular factors including microtubule regulatory proteins, neurotransmitters, glucocorticoids, and growth factors. Dendritic spines, which represent an important component of the synapses between hippocampal neurons, undergo structural plasticity in terms of formation, shedding, expansion, and atrophy. The potentiation or depression of synaptic activity, modifications in the synaptic function or number of synapses, and the generation of new dendritic spines facilitate the integration of neurons into networks [4, 6].

Functional plasticity refers to the regulation of neurotransmission, the reorganization of synaptic components and receptors, and the regulation of the strength or efficiency of synaptic transmission. Synaptic plasticity can manifest as the growth of new synaptic connections or pruning of existing ones, as changes in the strength or efficacy of synaptic transmission, and as modulated excitability of existing synapses. There are many forms of synaptic plasticity, such as post-tetanic potentiation (PTP), long-term potentiation (LTP), and long-term depression (LTD) [4].

The interaction of structural and functional plasticity is evidenced by synaptic plasticity leading to structural modifications in dendritic spines. The cellular basis of these adaptations is thought to be synaptic plasticity leading to changes in the functional connectivity of neuronal networks in the brain. The remodeling of synaptic connections can be measured both at the functional and at the morphological level. The function of synapses can be evaluated through electrophysiological recordings of spontaneous activity and can involve the release of neurotransmitters. The morphology of synapses refers to the number and shape of contacts between neurons. These changes can be accompanied by variations in the shape, number, and function of the other cells that surround the neurons, including glia, endothelial cells, and resident immune cells such as microglia and perivascular circulating macrophages [7].

2.2 Modulation of plasticity

Neuromodulation controls synaptic plasticity at different levels: at the network level by directing the flow of information, at the circuit level through changes in the balance of excitation and inhibition, and at the synaptic level through the modulation of intracellular signaling cascades [6]. The modulation of structural plasticity is influenced by various molecular, physiological, or external factors.

Molecules such as vascular endothelial growth factor (VEGF) can promote neurogenesis, and hormones and peptides involved in glucose homeostasis also support neurogenesis by influencing the proliferation, differentiation, and survival of NSCs. Physical activity, exposure to an environment rich in sensory stimulation, learning (especially task-specific learning), exploration, and social interactions have all been shown to significantly facilitate neurogenesis. However, neurogenesis declines dramatically with age or under intense stress. Antidepressant treatment, on the other hand, can improve hippocampal neurogenesis through its positive effects on mood and cognition [2, 3].

2.2.1 The role of glucose metabolism in the modulation of hippocampal plasticity

The activity of neural circuits in the hippocampus is influenced by metabolic stimuli and energy supply. Hormones and peptides involved in glucose homeostasis, such as insulin, glucagon-like peptide-1 (GLP-1), insulin-like growth factor-I (IGF-I), and ghrelin, are thought to play a role in neural plasticity, including synaptic plasticity (i.e., synapse potentiation and depression), structural plasticity (i.e., dynamics of dendritic spines), and adult neurogenesis, thus, impacting cognitive performance. For example, insulin and IGF-I activate the phosphatidylinositol trisphosphate kinase (PI3K)/Akt and Ras-mitogen-activated protein kinase (Ras/MAPK-ERK) pathways, which affect gene expression and have a significant impact on NSC proliferation and neuronal activity. GLP-1, secreted by the gut in response to satiation, helps to regulate glucose homeostasis by enhancing insulin release from pancreatic β-cells and increasing glucose sensitivity. In contrast, ghrelin, secreted by the stomach, stimulates feeding behavior and counteracts the effects of insulin.

However, it's important to note that chronic hyperstimulation of insulin/IGF-I effectors can lead to a premature decrease in the NSC pool, thus insulin may exert either beneficial or detrimental effects on NSCs depending on the timing and duration of stimulation [4].

2.2.2 The role of physical exercise and particularly running in hippocampal plasticity

The hippocampus demonstrates significant functional plasticity in response to exercise. Specifically, running increases neurogenesis in the DG of the hippocampus, leading to improved synaptic plasticity and memory function, particularly in spatial memory and pattern separation. Neurotrophins, neurotransmitters, and other peripheral components, such as myokines, hepatokines, and adipokines, may mediate these changes in neural plasticity in response to physical exercise [8].

The process of establishing connections between developing neurons and various brain areas involves multiple neurotransmitters. Glutamate and gamma-aminobutyric acid (GABA) serve as the primary excitatory and inhibitory neurotransmitters, respectively, in the brain. Both play a role in regulating the integration and survival of new neurons. Glutamate is particularly important for exercise-induced changes in DG synaptic plasticity, as running enhances DG LTP, a form of synaptic plasticity that is thought to be a cellular model for learning and memory [5].

Neurotrophins, such as brain-derived neurotrophic factor (BDNF), play a significant role in brain plasticity. BDNF expression levels in the hippocampus are increased by running in animal models, leading to improved hippocampal plasticity, spatial memory, and object recognition. In humans, exercise-induced increases in BDNF serum levels have been linked to changes in hippocampal volume, while reduced BDNF levels are associated with age-related decline in hippocampal volume. BDNF promotes synaptic plasticity through downstream targets such as cAMP-response element binding (CREB) protein, synapsin I, and synaptophysin, while simultaneously increasing its own messenger RNA (mRNA) and its receptor tyrosine kinase B (TrkB) [5].

Exercise also modulates cerebrovasculature, leading to better perfusion and delivery of oxygen, nutrients, neurotrophins, and other factors that promote brain function. Walking and running increase cerebral blood flow (CBF) in multiple regions within the brain, including the hippocampus. Additionally, skeletal muscle releases myokines that appear to play a role in neural plasticity as well [5].

2.2.3 The role of rich environments in neural plasticity

The impact of the living environment on both overall health and brain plasticity is significant. Prolonged stress can increase the risk of various diseases, including cardiovascular diseases, cancers, neuropsychiatric disorders, and neurodegenerative diseases. In contrast, a stimulating environment can contribute to improved health and behavioral performance by optimizing brain plasticity. In depressed individuals, a decrease in hippocampal volume has been observed. Conversely, a stimulating environment, such as high-level spatial orientation training, has been associated with an increase in hippocampal volume [8].

The ability of an enriched environment to stimulate hippocampal growth can be explained by an increase in the density of dendritic arborization, the length and volume of myelinated fibers, and the number of dendritic spines in the hippocampus [7]. At the molecular level, an environment rich in social interaction, learning, exploration, and sensory stimulation promotes the expression of neurotrophic factors in the hippocampus. These factors, including BDNF, IGF-1, and NGF, can affect neurogenesis and synaptic plasticity by exerting their influence on various cell types. In astrocytes, they regulate metabolism, recycle, and eliminate metabolites, promote myelination in oligodendrocytes, regulate synaptic pruning in microglia, and enhance angiogenesis in endothelial cells. Increased neuronal activity intensifies the release of neurotrophic factors, which in turn, boosts neurogenesis and synaptogenesis, encouraging further neuronal activity [7]. The immune system plays a role in normal behavioral processes and circulating T cells have a supportive role in brain function, cognitive abilities, and hippocampus neurogenesis, although the mechanisms of their involvement in hippocampal plasticity are still not fully understood [7].

In nonpathological conditions, T cells do not have direct access to the brain parenchyma, but a small number of T cells are present in the brain's blood vessels, in the choroid plexus, and in the meninges. T cells are thought to interact directly with endothelial or epithelial cells of the choroid plexus or contribute indirectly to the release of factors such as cytokines or chemokines in the blood or cerebrospinal fluid. Alternatively, T cells could also participate in the functioning of hormonal systems that regulate brain plasticity [7, 8].

2.2.4 The effects of stress on neuroplasticity

Transient mild stress can enhance learning and memory, but chronic or severe stress leads to the disruption of hippocampus-dependent memory. Extended or high-dose treatment with glucocorticoids seems to have a similar effect, as patients treated with glucocorticoids have presented impairments in hippocampus-dependent memory [9]. The hippocampus can be damaged at the level of morphological neuroplasticity by sustained levels of stress or glucocorticoids due to the atrophy and retraction of the apical dendrites of hippocampal pyramidal cells. Prolonged intake of corticosterone in higher doses than are typically achieved in vivo can even result in the death of hippocampal pyramidal cells [8].

Many forms of synaptic potentiation are triggered by increases in synaptic calcium influx and the local concentration of the second messenger molecule cyclic AMP (cAMP). Local calcium influx is activated only when presynaptic and postsynaptic cells are depolarized simultaneously, while cAMP is regulated by many modulatory neurotransmitters, including serotonin, dopamine, and norepinephrine, as well as by calcium. Local elevations in calcium and cAMP induce events required for short-term synaptic plasticity. Inducible transcription factors and cAMP- CREB turn on effector genes that contribute to the stabilization of synaptic plasticity [10]. Brain-derived neurotrophic factor (BDNF) influences synaptic plasticity in both the presynaptic and postsynaptic cells. Other growth factors have also been demonstrated to influence

LTP, including the VEGF. This has been implicated in the actions of stress and antidepressant treatments, like BDNF. Acute and chronic stress lead to reductions in hippocampal BDNF mRNA levels, suggesting an impairment of some of the mechanisms of neuroplasticity [10].

Other growth factors are regulated by stress, nerve growth factor (NGF) is upregulated by chronic stress, while VEGF is suppressed by chronic stress. Similarly, the angiogenic actions of VEGF are impaired by glucocorticoid treatment. The involvement of trophic factors beyond BDNF suggests that a multifaceted apparatus of neuronal support may be undermined by stress and, possibly, enhanced by antidepressant therapies [10, 11]. Chronic stress or elevations in glucocorticoids can lead to neuronal atrophy, especially dendritic retraction in cells of the CA3 cell field. Glucocorticoid excess increases glutamate release in the CA1 region of the hippocampus, and chronic behavioral stress increases extracellular levels of glutamate in the CA3 region. Glutamate antagonists can attenuate or block some of the effects of chronic glucocorticoid excess on dendritic morphology in the hippocampus [11].

2.2.5 The role of antidepressants in the modulation of neuroplasticity

When depression leads to a deficit in neuroplasticity, then antidepressant treatments may enhance neuroplasticity and even reverse deficits produced during the symptomatic period. In addition to enhancing functional neuroplasticity, antidepressants produce structural plasticity. This is observed at several different levels, including the numbers of synapses, spines, dendrites, and even the number of cells themselves.

Synaptic potentiation and other forms of neuroplasticity are controlled by both positive and negative regulatory mechanisms. Under certain circumstances, LTD may provide a homeostatic counterbalance to excessive synaptic potentiation. Inhibitors of the signal transduction cascades that contribute to synaptic potentiation also provide a counterbalancing influence. Examples include phosphatases such as calcineurin, which can antagonize signaling through the MAPK cascade, and phosphodiesterases, which break down cAMP and thus attenuate PKA (protein kinase A) mediated signaling and other cAMP-dependent processes [10, 12].

Understanding the properties of adult neurogenesis may contribute to enhancing hippocampal functions. However, it is unknown whether increasing adult hippocampal neurogenesis is sufficient to improve cognition and mood. Furthermore, the stimulation of adult hippocampal neurogenesis, when combined with an intervention such as voluntary exercise, produces a robust increase in exploratory behavior. In contrast, increasing adult hippocampal neurogenesis on its own does not produce an anxiolytic or antidepressant-like behavioral response [13].

As mentioned before, the DG subregion of the hippocampus is a substrate for both cognition and mood regulation. Evidence from neuroanatomical, computational, electrophysiological, behavioral, and human brain imaging studies converges to suggest the crucial role of the DG in the formation of new episodic memories by transforming similar experiences or events into discrete, non-overlapping representations, a process known as "pattern separation." Consistent with these DG functions, the ablation of adult hippocampal neurogenesis impairs pattern separation and blocks some of the behavioral effects of antidepressants. In sharp contrast, the impact of selectively increasing adult hippocampal neurogenesis on cognition and mood is not known. Addressing these questions has proven difficult owing to a lack of strategies that selectively increase adult neurogenesis [14].

3. Connectivity

The hippocampus is a complex structure of the brain responsible for multiple functions, such as learning, memory, and emotion, all of which we rely heavily on and value greatly in our daily lives. Understanding its role in the organization of the entire brain and especially the dynamic relation of its different subregions with the cerebral cortex is of utmost importance to discern the mechanisms of brain disorders, such as Alzheimer's disease (AD), temporal lobe epilepsy, and schizophrenia.

The capacity to store memories is established during infancy and is related to sleep, specifically slow-wave activity (SWA). This supports the brain's processing of information in two ways. First, information encoded during wakefulness is mediated through sleep, renewing the brain's capacity to store new information. Second, sleep supports the creation and strengthening of long-term memory [15]. These functions rely on the circuits that connect the hippocampus and the prefrontal cortical networks, helping to store experienced events into new memories. These effects are observed primarily during deep, non-rapid eye movement sleep (NREM) and, specifically, slow-wave sleep (SWS), a deep form of NREM.

Unlike adults, infants and children have a greater need for sleep, spending almost half of their life sleeping in a disorganized manner at first, with periods of daytime sleep, gradually shifting to less, but more organized night sleep, with established slow oscillations (SOs). Studies have shown that sleep disturbances in children are associated with learning difficulties [16]. This highlights the critical nature of this period in brain development. Practically, sleep time decreases gradually with growth, but sleep patterns change, with deep sleep defined by an increase in SWA, comprising 75% of total sleep time by the age of 2–3, compared to the first post-natal months, when there is an equal distribution between REM (the so-called "active sleep" because of the presence of muscle impulses) and NREM (also known as "quiet sleep"). Spindle activity, defined as NREM oscillations resulting from waxing and waning patterns in neuronal firing, is also relevant for cerebral development and follows the same course as SWA, showing a significant increase during childhood and puberty and reaching a plateau during adolescence. The consistent increase in SWA and the reaching of optimal spindle density, which occurs until the onset of puberty, promote a significant enhancement of synaptic network connectivity. It is important to keep in mind that the inherent maturation of synaptic connectivity that comes with age must be further sustained by utilizing the network; the network connectivity grows stronger with encoding information [15].

The episodic representation of events in time and space is based on the activity of the hippocampus and the prefrontal cortex. On the other hand, semantic and procedural memories like recalling memorized facts or how to ride a bike, for example, processes that involve repeated training, do not necessarily involve hippocampal function. The postnatal hippocampus reveals a complex dynamic as it attains a number and density of synapses like those in adults as early as first six months of life. However, infants present limited memory capacity, with hippocampus-dependent episodic memory performance increasing slowly over time. This leads to a higher rate of forgetting in children compared to adults, and it could explain the absence of autobiographical memories before three years old. Following this initial three-year period, there is a relevant increase in SWA and a shift of sleep activity toward more frontal cerebral areas during childhood, gradually enhancing the ability to store more information for a longer time, thus consistently expanding the episodic memory capacity up until the early stages of puberty. Inside this time window, the synaptic connectivity within the cortical networks undergoes significant transformation, leading to an overall evolution in memory capacity and, together with it, an intensified synaptic plasticity [15].

Once the transition to adolescence begins, the human brain develops new abilities and progressively reorganizes itself over time, increasing cognitive functioning and memory performance. Puberty and adolescence bring a reduction of sleep time and sleep pressure during wakefulness, together with the stabilization in SWA. If a first stream of gonadal hormones initiates the definition and the structure of neuronal networks during childhood, a second stream of the same hormones is set to completely grow and trigger those networks during puberty [17]. It is in this period when the different regions of the brain work in increasing synchrony, and the brain becomes a system within which all its constituent elements (brain areas) impact and influence one another, in a functional interaction that can be evaluated in terms of functionality and effectiveness. Functional connectivity is characterized by temporal correlations between spatially dispersed neurophysiological incidents [18]. Of course, effective connectivity is the one offering a more complex overview of the interconnection and exchange within the cerebral network [19]. Research on this topic revealed that there is a remarkable development of memory processing capacity provided by maturation, that memory performance is enriched with age, evolving from neuronal connections based on elementary visual processing to connections governed by multi-modal processing on a higher level. During adolescence, there is a great development in white matter pathways, an essential element for efficient connectivity within the cerebral network, gradually translating into more substantial, more resilient, and more efficient structural connectivity [19].

During adolescence, the brain undergoes significant development and reorganization. One aspect of this development is the increased influence of sex on cerebral processes and changes within the hippocampal network. Research has shown that adolescent girls reach peak brain volumes earlier than boys, while testosterone is associated with a greater increase in white matter in boys [20, 21].

As individuals transition into adulthood, cognitive processes become more complex and problem-solving becomes increasingly reliant on memory. Additionally, there is a notable improvement in pattern separation, creativity, conceptualization, higher-order self-observation, perspective, future thinking, and risk-taking. These abilities are all connected to cerebral mechanisms that control emotion and motivation. Emotion can also play a role in functional connectivity and decision-making, with trait anxiety found to impact decision-making and risk-taking. Research has also revealed that when people are faced with risky situations, two functions are activated in decision-making: evaluation and choice. The first involves assessing various possibilities, while the latter involves preferring one option over others for reasons deemed pertinent to the situation [22, 23].

A crucial aspect of decision-making is weighing the potential consequences of a choice. The anticipation of negative consequences, such as loss or a negative perception of risk, generates aversion as a behavioral predisposition. Studies have found that aversion to loss and risk triggers activity in cerebral areas related to motivation and emotion, including the amygdala and the prefrontal cortex [24]. Trait anxiety has been shown to undermine decision-making in risky situations by impacting episodic future thinking and episodic memory, cognitive processes related to the hippocampus. It is also associated with contextual fear learning [25] and is thought to be a part of contextualization and episodic prospective memory. Individuals affected by trait anxiety show a deficit in attentional control, tending to focus on negative outcomes, which can lead to a reduction in rational cognitive function that relies on executive control. This can result in a preference for instant rewards over delayed, greater gratification, which suggests how trait anxiety interferes with the functional connectivity between the hippocampus and other cerebral areas. Conversely, when not disturbed by impairments such as anxiety or cognitive bias, the functional connectivity between the hippocampus and connected cerebral regions can enable creativity. Connectivity of the hippocampus with the medial temporal lobe leads to the creation of new associations starting from familiar concepts, while the connectivity of the middle temporal gyrus (MTG) with the hippocampus enables the construction of new concepts or categories and its connectivity with the executive control system promotes breaking the limits of the old concepts and patterns.

Low-quality sleep has been associated with depression and its detrimental effects. Studies have revealed that the lateral orbitofrontal cortex is affected by depression, being sensitive to the areas triggered by the failure to attain anticipated rewards. The functional connectivity between this cortical area and the precuneus, a cerebral area related to the perception of self, generates low self-esteem by linking this self-representation with the previously mentioned lack of rewarding, activating depression symptoms and causing distressing incessant thinking which has a heavy, negative impact on sleep [26].

With age, significant changes take place in the hippocampus and its subfields, with relevant effects on memory, cognitive and functional connectivity. Studies have shown that certain hippocampal regions function differently depending on the type of information being processed and the cognitive processes involved. For example, retrieval-based mental operations that process visual, spatial, or neutral information appear to rely on the hippocampal posterior regions, whereas encoding- and association-based mental operations that process emotional or motivating information appear to depend on the hippocampal anterior regions [27]. Additionally, volumetric declines in the DG/CA3 and CA1 regions have been reported in several studies [28].

Aging seems to make the functional dissociation between the anterior and posterior hippocampus more pronounced in terms of how each contributes to memory processing. Studies have also observed a gradual shift from one to the other over time. Some research has shown an increasing dominance of the connectivity involving the posterior hippocampus in aging individuals, suggesting that the brain undergoes a reorganization of cerebral networks and their connections as a means of sustaining cognitive processes. This may be an adaptation to the fact that the anterior hippocampus is the first to suffer from atrophy [29]. However, other studies have found that the volume and proportions of the anterior and posterior hippocampus regions were not significantly different in aging subjects compared to young individuals. However, a general decrease in the functional connectivity between the hippocampus and the rest of the brain as people age is typically observed, with several cerebral areas developing further functional connectivity to the posterior hippocampus. This may be related to the additional effort made by an aging brain to maintain normal memory function [29, 30].

The brain is a highly complex network, with a wide range of subregions, each of them essential for normal cognitive functioning. Connectivity, the continuous flow of information and influence of one subsystem (a cerebral area) on others, is vital for sustaining the brain's complex capacity for memory and learning, the very foundation of human consciousness. Connectivity is formed during infancy and defined by continuous adjustments throughout life, undergoing changes specific to each life stage, and finally adapting to aging and related cognitive decline.

4. Pharmacological interventions

Although the hippocampus is the most widely recognized brain region linked to memory and learning, it is a vulnerable, plastic structure that can be affected by a wide range of triggers and stimuli, such as pharmaceutical elements. Due to its fairly straightforward structural connectivity and its different roles and functions, the hippocampus is an ideal model for pharmacological research. Numerous medications from several pharmaceutical classes have been demonstrated to have favorable effects on its function, while others were shown to elicit harm.

4.1 Prescription medication

4.1.1 Lithium

In addition to its well-known therapeutic effects on bipolar disorder and depression, lithium has neuroprotective effects on neurodegenerative diseases like traumatic brain injury (TBI) [31]. Lithium has also been shown to have neuroprotective properties. A wide range of neurotrophic effects is displayed by lithium. For instance, lithium enhances LTP and hippocampal neurogenesis, and elevates cytoprotective B-cell lymphoma protein-2 (bcl-2) levels in a number of rodent brain regions and in cultured cells [32–34]. Due to this, neuroprotective and antiapoptotic properties are produced. In people with bipolar disorder, lithium builds up in the brain's neurogenic regions and has a positive impact on hippocampal volume [35]. However, prolonged use of lithium, especially at low doses, results in decreased neurogenesis and hippocampal atrophy [31].

4.1.2 Estrogens

Beyond their role in regulating reproductive function, estrogens have numerous additional effects on the nervous system. According to reports, they have an impact on verbal fluency, verbal memory tests, spatial task performance, fine motor skills, signs of Parkinson's disease, and tardive dyskinesia [36]. The neuroprotective properties of estrogens are well known. Sex hormones have the hippocampal region as a target, and large amounts of estrogen are produced by hippocampal neurons [37]. It has been established that ovarian estrogens affect neurogenesis in the DG. Also, estrogen has a variable impact on inflammatory markers and a considerable impact on thrombosis and thrombolysis. It also leads to higher CBF, glucose transport, and glucose metabolism [38]. However, a few of the effects of estrogen may be negative. Some estrogens' prothrombotic properties may have harmful effects on the cerebral vasculature, and proinflammatory effects may be damaging [39].

4.1.3 Sodium valproate

After more than 40 years of use, sodium valproate has a good safety profile. Its broad activity against both generalized and partial seizures, ability to stabilize mood in bipolar disorder, and effectiveness in treating migraines make it unusual among anti-convulsants. When compared to other anticonvulsants, it has comparatively few side effects and is frequently used by epileptic patients, sometimes with great success for decades [40]. Sodium valproate has neuroprotective properties. It enhances cytoprotective protein bcl-2 in the central nervous system, encourages neurite outgrowth, and activates the extracellular signal-regulated erk pathway, a signaling pathway used by numerous endogenous neurotrophic factors to control neurogenesis, neurite outgrowth, and neuronal survival [41]. Additionally, in the DG of the hippocampus, valproate stimulates neurogenesis possibly via the ERK pathway and guards against excitotoxicity [42].

4.1.4 Fluoxetine

Fluoxetine is a selective serotonin reuptake inhibitor (SSRI) with therapeutic indications for bulimia nervosa, posttraumatic stress disorder, anxiety disorders, and major depressive disorder [43]. One of the biological mechanisms of fluoxetine in the treatment of depressive disorders is suggested to be the way it regulates dendrite atrophy of hippocampal neurons caused by NOS overexpression, by halting the overexpression of nitric oxide synthase [44]. Moreover, it overturns cell proliferation in response to unavoidable stress [45]. It has been discovered that an increment in hippocampal S100beta content is the cause of fluoxetine's neurotrophic action [46]. The stimulation of neurogenesis in the hippocampus may serve as a bridge between the behavioral effects of ongoing antidepressant therapy, such as fluoxetine [47]. In depressed rats, fluoxetine can boost the number of hippocampal neurons, enhance the stereology of synaptic structures, and restore the structure and functionality of the hippocampus [48].

4.1.5 Phenytoin

Phenytoin is an anti-seizure medication that has been studied in clinical trials for nearly eight decades. It is mostly used to treat tonic-clonic and partial seizures [49]. The motor region of the cerebral cortex is subject to a highly selective inhibitory effect from phenytoin. Phenytoin extends the neuronal refractory period by binding to the inactivated state of the $Na+$ channel as its mode of action [50]. Phenytoin prevents neurodegeneration by blocking the effects of chronic stress on the number of pyramidal neurons in the hippocampal CA3 area and the length of all their apical dendrites [51]. It also prevents dendritic atrophy and inhibits glutamate release [52]. Phenytoin has several additional advantageous neuroprotective properties, for instance, increases right brain volume by 6% in PTSD and has antiseizure properties [53, 54].

4.1.6 Cerebrolysin

Cerebrolysin, an amino acid and peptide blend, mimics the biological effects of neurotrophic factors and has been shown to have positive effects when given to patients who have experienced an ischemic stroke while exhibiting a benign safety profile [55]. The mechanism of action of Cerebrolysin is multi-faceted and has been demonstrated in a range of studies, including animal and *in vitro* studies. One of the key mechanisms of action of Cerebrolysin is its impact on the hippocampus, which is a crucial area for learning and memory. By regulating the growth of synapses and promoting neuroplasticity in the hippocampus, Cerebrolysin has been shown to improve memory and cognitive function in patients with ischemic stroke [55]. Additionally, Cerebrolysin activates the Sonic Hedgehog pathway, which boosts neurogenesis and oligodendrogenesis in the hippocampus, contributing to the development and structuring of the brain [56]. Cerebrolysin has been shown to reduce the production of free radicals and programmed cell death, control the inflammatory response, and reduce the toxic effects of neurotransmitters (excitotoxicity) [55]. Cerebrolysin promotes neuroplasticity, neurogenesis, and oligodendrogenesis in the hippocampus, and reducing oxidative stress and inflammation to promote neuroprotection and neurorecovery. The biological agent, therefore, has the potential to improve cognitive function and reduce the risk of damage after brain lesions.

4.2 Drug use

Prescription medication used for diverse purposes can impact the hippocampus in varied ways, but it is also important to look at the way addictive substances affect it. Drug abuse and addiction are major global health issues, and both are frequently associated with various types of neurological impairment. Due to the hippocampus' high level of plasticity and its capacity to support declarative and contextual memories, drugs may cause changes in hippocampal function that have a significant impact on behavior. Drug-induced changes to hippocampus-dependent learning and memory appear to play a significant role in the emergence and maintenance of drug addiction, according to a large body of human and animal studies [57].

4.2.1 Cocaine

Cocaine use is typically characterized by compulsive behavior and obsessive drug seeking. It is a highly addictive psychostimulant derived from the leaf of the Erythroxylon coca plant [58]. It is the most prevalent illegal substance among patients who seek emergency care and has detrimental effects on the heart, lungs, and mind [59–61]. Cocaine directly affects several brain areas in the mesolimbic circuit, including reward-related areas such as the nucleus accumbens and ventral tegmental area, as well as areas that govern cognition like the prefrontal cortex and hippocampus [62]. Given its involvement in learning and memory, as well as reward, the hippocampus may be a key area for the rewarding effects of cocaine [58, 63]. Research has shown that the ventral hippocampus mediates cue-induced and cocaine-primed reinstatement of cocaine self-administration, in contrast to the dorsal hippocampus [64, 65]. Additionally, the dorsal and ventral hippocampus may play different roles in stress-related relapse to cocaine seeking. Studies have shown that stress has opposing effects on these regions' synaptic plasticity, with stress increasing ventral hippocampal synaptic plasticity while decreasing synaptic plasticity in the dorsal hippocampus [66]. Therefore, during a time of high stress such as cocaine withdrawal, the ventral hippocampus may play a larger role in the resumption of cocaine-seeking behavior. Furthermore, cocaine CPP has been found to increase the phosphorylation of hippocampal cAMP-response element binding protein (CREB), a transcription factor that is crucial for the formation of long-term memory and synaptic plasticity [67, 68].

4.2.2 Amphetamine

Racemic a-methylphenethylamine (amphetamine, also known as "speed") was first discovered in 1910. It was later synthesized and sold under the trade name Benzedrine to treat a variety of ailments, including narcolepsy, depression, Parkinson's disease, and pulmonary dysfunction [69]. However, the euphoric effects of amphetamine were quickly recognized, leading to its abuse for its pleasurable effects, such as feelings of self-confidence, energy, and alertness [70].

Research soon showed that amphetamine and its derivatives, such as dextroamphetamine, have procognitive effects, including improved intelligence, concentration, and intellectual performance [71, 72]. These effects were later supported by studies that showed that amphetamine and its derivatives enhance memory consolidation [73], memory recall [74], attention, and psychomotor performance [75, 76], information processing [77], and logical reasoning [75].

Amphetamine was initially used to treat ADHD due to its attention-enhancing properties, but later, medications with fewer psychoactive side effects replaced it [78]. However, despite its acute procognitive effects, amphetamine and methamphetamine use has detrimental consequences for cognition, including impaired memory, attention, flexibility, inhibition, and decision-making. Additionally, its toxic impact leads to the degeneration of neuronal apoptosis and monoaminergic terminals [79].

4.2.3 Cannabis

Cannabis, a plant that has been used for therapeutic, recreational, and commercial purposes for many years, has several products derived from it, such as hemp oil and hemp fiber. Hashish and marijuana are examples of recreational drugs made from cannabis subspecies that have been selectively bred to produce a high yield of D9-tetrahydrocannabinol (THC), the plant's primary psychoactive component. Studies have shown that the administration of THC and cannabidiol impairs contextual fear conditioning and hippocampus-dependent spatial learning in the Morris water maze and radial arm maze [80, 81]. This is similar to the effects of opiates but different from the effects of cocaine and amphetamine [82]. Cannabinoids have been shown to impair all phases of memory processing, including encoding, consolidation, and retrieval. The amnestic effects of cannabinoids have been attributed to several mechanisms, including effects on LTP and LTD, as well as the suppression of neurotransmitter (GABA, glutamate, acetyl choline, and dopamine) release [83]. Hippocampal plasticity deficits from acute cannabis use may be the cause of the disruption of hippocampus-dependent learning. For instance, in the hippocampus, acute THC decreased the amplitude of both spontaneous and conditioned stimulus-evoked potentials [84].

4.2.4 Opiates

Opiates are a group of psychoactive substances that are either synthetically made or naturally derived from the opium poppy. They include morphine, diacetylmorphine (also known as heroin or diamorphine), codeine, oxycodone, and methadone. Patients may receive a legal prescription for opiates such as oxycodone for pain relief. However, as tolerance grows, these legally prescribed opiates may lose their analgesic effects, which can lead to drug dependence as users attempt to increase their effectiveness or prevent withdrawal effects [57]. The consequences of opioid exposure on hippocampus anatomy and hippocampus-dependent learning and memory imply that some of the cognitive dysfunction seen in opiate abusers may be related to altered hippocampal function. Despite this, several studies have suggested that opiate abuse-related cognitive decline may also be linked to compromised frontal lobe function [85, 86]. For example, opioids have been shown to reduce adult neurogenesis in the hippocampus [87] and alter clathrin, a protein that is linked to the density of synapses in the hippocampal nucleus [88]. Furthermore, there is evidence that chronic use of heroin and morphine harms hippocampus-dependent spatial learning in the Morris water maze, radial arm maze, and Y-maze [89].

5. Non-pharmacological interventions

5.1 The importance of sleep

The strong connection between sleep and neurological health is well-established, and interest in this subject is increasing. The natural process of aging brings about alterations in sleep patterns, which can further lead to neurodegenerative disorders. Finding viable solutions for confronting cognitive impairment has been a growing interest among researchers in recent years; as a result, focusing on sleep enhancement methods has become an important path to be followed in order to achieve this goal.

It has been observed that older adults are commonly affected by sleep fragmentation and a significant reduction in SWS [90], together with a reduction in SWA and sleep slow oscillations (SOs). These factors have a negative impact on sleep-mediated memory retention [91]. These factors are relevant for approaching neurodegenerative disorders because early intervention in sleep-wake calibration has been suggested as a potential factor in delaying or alleviating such conditions, such as Alzheimer's disease (AD), where the focus is increasingly shifting to the early stages [92], using neuromodulation as a strategy for delaying the disease progression as much as possible [93].

Memory consolidation depends on neuronal processes that take place during sleep when information related to declarative memories (the ones linked to remembering facts and events, and not skills) is transferred from the hippocampus to the neocortex and integrated into long-term existing information [94]; the synchronization of different networks within the neuronal system during three essential waveforms (hippocampal sharp-wave ripples, SOs, and spindles) of SWS is what the consolidation of sleep-dependent memory relies on [95]. Therefore, the solution for ensuring high-level functioning memory and cognition is represented by different interventions developed for enhancing sleep, especially SWS and SWA.

5.2 Transcranial electrical stimulation (TES)

Transcranial electrical stimulation (TES) is a non-invasive method that modulates neuronal activity through the application of moderate electrical current via scalp electrodes [96]. Studies have focused on both short-term and long-term stimulation (transcranial direct current stimulation—tDCS) during wake, nap, or nighttime sleep in healthy individuals and older adults with conditions such as mild cognitive impairment (MCI), Alzheimer's disease (AD), Parkinson's disease, multiple sclerosis, and stroke. Positive effects have also been observed in the treatment of psychiatric conditions such as schizophrenia and depression [97]. The impact of tDCS on the nervous system is related to its ability to modulate neuronal excitability in polar coordinate systems. Anodal tDCS has been shown to increase excitability, while cathodal tDCS reduces it, producing lasting effects on the subjects' motor cortex. This type of system has also been suggested to affect synaptic plasticity in the hippocampus, specifically related to LTP and LTD. However, the data indicate that the effects of tDCS on hippocampus plasticity are highly dependent on current polarity and the specific subregion studied. While the results are not homogenous and linear, the primary advantage of nap studies is that tDCS has been observed to enhance sleep, and as a result, sleep-mediated declarative memory.

More research is needed to understand the effects of tDCS during nighttime sleep and wakefulness [98]. Anodal tDCS has been shown to enhance memory through molecular adjustments in the hippocampal synaptic system. When applied before the acquisition step, tDCS is thought to increase proteins that enable glutamate signaling and ion channel activity, thereby strengthening memory performance [99].

5.3 Transcranial magnetic stimulation (TMS)

Transcranial magnetic stimulation (TMS) is a technique that uses a magnetic field to send electrical current pulses into the cerebral cortex, modulating cortical plasticity in both human and animal models [100]. Repetitive TMS (rTMS) has been shown to produce significant benefits for memory function in healthy young adults and to enhance cognitive functions and memory in patients with Alzheimer's disease (AD) and mild cognitive impairment (MCI). rTMS performed during sleep or immediately before sleep has been observed to increase sleep spindle-associated oscillations (SOs), positively impacting sleep-mediated memory [101]. Studies have also revealed the benefits of rTMS in enhancing spatial memory, due to its ability to calibrate and sustain hippocampal synaptic plasticity. Low-intensity magnetic stimulation (LIMS) has been shown to enhance synaptic activity and stimulate nerve regeneration and cell morphology maintenance, while high-intensity magnetic stimulation (HIMS) is thought to reduce synaptic plasticity [102]. The most common areas targeted for rTMS are the ones related to the frontal association (dorsolateral prefrontal cortex—DLPFC), for its impact on several cognitive functions such as working memory, its important role in treating psychiatric conditions like severe depression and for being widely considered a safe region for performing rTMS. Left and right DLPFC seems to respond differently to rTMS. In previous research, rTMS to the left DLPFC has been linked to enhanced memory function, by increasing resting-state functional connectivity and brain activation patterns during encoding and retrieval in memory tasks [102]. That may be due to the prefrontal-mesiotemporal lobe circuitry potentiation and hippocampal synaptic multiplications [103]. The angular gyrus (AG) is also targeted less frequently, for its high functional and structural connectivity to the brain areas that support declarative memory functions. Both areas are viable targets for rTMS due to their optimal location directly under the skull, within the typical TMS structures [102]. Most procedures involving rTMS have focused on a single brain area at a time, but there have also been studies in which rTMS targeted multiple brain regions, and this type of stimulation has also been proven to be beneficial in modulating neuronal activity, suggesting a high potential for cognitive improvement. Research involving multitarget stimulation, in which several focal points in both frontal and parietal regions of the brain were simultaneously stimulated five days a week over a six-week period, showed significant improvements in AD patients regarding cognitive performance, lasting for up to 12 weeks, providing very encouraging results in the field and suggesting persistent modulation of the functional neuronal system [104].

5.4 Sensory stimulation

Sensory stimulation is a commonly used method for enhancing sleep and is represented by various techniques such as vestibular stimulation (induced through slow, gentle rocking, which has been observed to enhance SWA and the density of spindles in naps), olfactory stimulation (which is effective, especially when dealing with targeted memory reactivation), or auditory stimulation. Among these techniques, acoustic stimulation [105] has been found to be the most efficient. It is a safe, non-invasive method that can be applied even in a familiar environment by the beneficiaries themselves, but it is limited to those without any associated hearing impairment. The delivery of acoustic stimulation has been suggested to increase SWA and SOs, to show relevant effects on enhancing memory, and to also generate benefits beyond memory consolidation, such as improvements in cardiovascular and immune systems. Even if SOs originate in cortical and thalamic regions, they mirror global synchronous neural activity, that spreads throughout the neocortex and also in subcortical structures, like the hippocampus [106]. In addition to their involvement in synaptic downscaling and homeostasis, SOs are causal in the consolidation of memory. The coordination of fast-spindle activity (12–15 Hz) and hippocampus ripples to the up-depolarizing state appears to be crucial for this consolidating function, contributing, in this manner, to hippocampal plasticity [107]. The acoustic stimuli can be provided through headphones or even bone conduction. Some researchers have pointed out that unilateral stimulation can be as effective as bilateral stimulation in increasing SOs for both cerebral hemispheres. The most commonly used stimulus type in acoustic stimulation is pink noise, defined as one of the most prevalent noises in nature, which has a wide range of sound frequencies transferred in a 1/f distribution, with lower frequencies offering a more significant contribution [100]. Numerous studies have shown that acoustic stimulation using pink noise contributes to a lower complexity in neuronal activity during sleep, ensuring an increased balance in the sleep state resulting from optimal coordination of brain waves [100]. The methodology used for acoustic stimulation during sleep is also important because there is a fine line between beneficial stimulation for promoting slow waves and generating sleep fragmentation or arousal. This is why the search for optimal calibration of the stimulation is a continuous process, involving constant monitoring and adjustment. This calibration regards the intensity of sounds (the range present in most studies is 20–65 dB, either at a fixed or at a varied volume within these limits), timing (it has been observed that arbitrary acoustic stimulation or immediate delivery of stimulation following SOs lowest level stage do not contribute to SOs or sleep-mediated memory enhancement, but coordinating stimulation with the high levels of SOs); so a particular endogenous SOs time window, with the help of EEG monitoring during sleep, is shown to be essential for obtaining these benefits [106].

5.5 Physical exercise and diet

Physical exercise has been observed to be an effective means of preserving overall health and promoting optimal function of the nervous and cardiovascular systems. Additionally, it has been identified as a non-pharmacological intervention for hippocampal modulation, as it has been shown to have significant beneficial effects on cognitive and metabolic functions, as well as on cerebral plasticity. This is achieved through the reduction of metabolic-related disorders and inflammation in adipose tissue, as well as by contributing to the balance of insulin resistance [108]. In particular, for aging individuals who are susceptible to a decline in cognitive and memory performance, an active lifestyle and regular exercise routine can have a significant impact, including a reduction in the risk of dementia, which is expected to become a major public health concern in the coming decade [109]. Many studies, including those involving both healthy adults and those with neurodegenerative disorders, have reported significant cognitive benefits from regular physical exercise. The minimal frequency for achieving these benefits is suggested to be at least 30 minutes per session, three times a week, for a minimum of 16–24 weeks [110].

There have been several attempts to correlate the intensity of exercise to cognitive performance, but the measurement of memory capacity is too complex for a linear and clear cause-effect model to be identified. Most studies have revealed that different intensity levels in exercises engage and enhance different memory processes, and a determinant factor in cognitive performance is optimal recovery post-exercise. High-intensity physical exercise has been observed to affect cognitive processes and complex assignments during training or immediately after it, difficulties in verbal acquisition and focus have been noted, but can have beneficial effects on working memory enhancement after proper rest when fatigue does not weigh in the equation of cognitive performance. There are outcomes pointing to both aerobic exercise promoting improvement in executive memory skills and to more moderate exercise boosting overall memory performance, so the optimal approach would be combining and alternating the two [111]. Running has been proven to lower hippocampal basic metabolic panel (BMP) levels, and transgenic mice with decreased BMP signaling showed impressive improvements in hippocampal cognitive function and neurogenesis [112].

The possible mechanisms behind linking memory improvement to exercise are multiple and on varied levels and are yet to be fully identified and assessed. It has been observed that an exercise routine enhances cerebral blood circulation, especially within the hippocampal area, which is mainly associated with memory and learning processes. The assumption is that this enriched blood circulation is a result of lactate, secreted by the skeletal muscles when contraction occurs and then used in different cerebral areas for optimal blood pumping and oxygenation and for boosting cerebral metabolism [111]. Lactate has also been associated with neurogenesis in some studies, being thought to be a contributor to increased cerebral gray matter integrity and reduced atrophy in the hippocampus, with a strong impact on memory and learning capacity [113]. Additionally, contraction of skeletal muscles has been shown to promote mitochondrial function, with a significant direct impact on resistance to oxidative stress, metabolism, cell survival, and proliferation of energy-productive proteins [114], but it was discovered that acute exercise can lead to mitochondria having a negative impact on memory performance by producing reactive oxygen species (ROS). This is why an exercise is a beneficial tool for neuronal processes only when approached regularly, at moderate levels. Furthermore, the contraction of the skeletal muscles during exercise seems to secrete several proteins that generate an increase in neurotrophic factors, which promote and sustain long-term memory and plasticity within the hippocampus [111]. The endocannabinoid system (ECS) is a biological lipids network with an essential role in modulating the nervous system, by enhancing synaptic plasticity, neurogenesis, and other neurophysiological functions. The endocannabinoid (EC) has been revealed as an important factor in lowering cerebral inflammation, anxiety, and oxidative stress, so the fact that its levels are increased by physical exercise is another proof of the great potential exercise has in modulating neuronal plasticity and memory performance. Reduction of insulin resistance is another mechanism that relates physical exercise to improvements in hippocampal plasticity, as it has been observed to elevate the level of enzymes that increases insulin sensitivity and thus prevent neurodegeneration [111].

When it comes to diet, research has shown that it can have a direct and immediate effect on neuronal activity and efficiency. Studies have demonstrated a notable decline in the performance of hippocampus-reliable processes such as learning and memory after only four days of consuming a diet high in saturated fats and added sugars [115]. Recent literature on this subject also indicates that even a week of a high-fat diet can affect memory and learning capacity, causing difficulty in recognizing objects and leading to mood disruptions and impairments in hippocampal plasticity if this type of diet is continued for more than a week [115].

A high-fat diet can generate chronic inflammation, reducing the density of dendritic spines in granule neurons and altering LTP in the hippocampus. It can also increase the levels of ROS in body fat, triggering a neuroinflammatory response and damaging synaptic plasticity. This can activate metabolic syndromes, disrupting the secretion of inflammatory-regulator cytokines and having a significant impact on both peripheral metabolism and hippocampal microglial activation, resulting in significant damage to plasticity within the hippocampus [116]. Excessive ROS generated by hyperglycemia can also have a damaging effect on mitochondrial DNA and oxidative capacity, leading to a decrease in ATP formation and mitochondrial density within neurons and the emergence of insulin resistance, and reduced exercise endurance in skeletal muscles [117]. Neuronal plasticity can also be suppressed by a significant intake of palmitate, the most saturated fatty acid found in the cerebrospinal fluid and in circulation. Studies have shown that palmitate increases microglial activity, leading to lower insulin sensitivity and resulting in microglial inflammation, which can cause a serious decline in the growth of surrounding neurons and ultimately have damaging effects on learning and memory processes [116].

Caloric restriction has been identified as a feasible solution for cognitive and neurotrophic enhancement. It has the ability to regulate mitochondrial biogenesis, enabling synaptic plasticity within the hippocampus, thus, increasing memory and learning capacity. Intermittent fasting has also been shown to have a beneficial impact on hippocampus-reliant spatial memory function and on reducing seizure rate [116].

5.6 Environmental and lifestyle enrichment

The well-functioning of cerebral processes within the hippocampus can also be enhanced by environmental enrichment, as researchers have found in several studies. The hippocampus is the cerebral area that is most vulnerable to degeneration induced by aging, and its functional integrity and plasticity can be restored and maintained by multiple environmental stimulations on many and varied levels: cognitive, social, motor, and sensory. These stimulations are provided by continuously performing complex and challenging physical and mental activities, which are constituents of the paradigm called enriched environment (EE). In humans, exposure to a complex environment as early as possible is of high importance and has a lifelong impact on cognitive functions. Studies have revealed that people who benefit from elevated education are less likely to be affected by cognitive decline and the associated degenerative disorders [118]. EE was found to trigger certain changes within the hippocampal transcriptome, translating into lowered spatial learning deficits and other cognitive benefits. EE has also been observed to reduce anxiety levels and increase strength, specifically grip strength, as well as enhance muscle performance. The physical exercise performed in an EE is more stimulating and beneficial overall. The various stimuli offered by an EE trigger simultaneous signaling in multiple, different regions of the brain, facilitating learning and enabling the adaptive potential, thus maintaining a high level of cognitive functions and ensuring neuronal plasticity in the hippocampus.

For example, light has been shown to be an important factor that can influence neurogenesis, as it is related to sleep and the circadian rhythm. The body and brain can be negatively affected by the disruption of circadian rhythms, which can be caused by aging, neurodegenerative disorders, or harmful light regimes. Continuous light can impede hippocampal neurogenesis and cognitive function. Additionally, exposure to low light at night can decrease the expression of hippocampus neurotrophic factors. These findings are highly relevant to the negative effects of nighttime light, such as that emitted by electronic devices, on mental and cognitive function. Light has been proven to be the most effective trigger for synchronizing circadian rhythms with their surroundings. Importantly, rhythmic light and dark cycles regulate psychiatry and behavior even when there is no functional molecular clockwork present. As different suprachiasmatic nucleus (SCN)-dependent and independent pathways deliver different types of light information to the hippocampus, it becomes clear that light exposure and intensity can affect hippocampus-dependent learning and memory. This effect is likely caused by an increase in active p21-activated kinase 1 (PAK1) and an increase in CA1 LTP in the hippocampus. However, it is unclear whether rhythmic light and dark conditions, as opposed to persistently dark conditions that do not change with the circadian cycle, promote neurogenesis [119].

The non-pharmacological factors that have been observed to modulate hippocampal plasticity are non-invasive methods adopted to improve memory and learning processes. Transcranial electrical and magnetic stimulation, which involves the use of scalp electrodes to send moderate electrical impulses to the brain, has been shown to have beneficial effects on sleep-dependent memory enhancement and cognitive function improvement in patients with neurodegenerative disorders. Sensory stimulation is another approach that can be used to enhance memory and learning performances, with acoustic stimulation using pink noise proving to be the most efficient. Adopting a ketogenic diet, practicing intermittent fasting, and engaging in physical exercise have also been shown to be beneficial in promoting learning processes and memory functional enhancement for not only overweight individuals but also for the general population, as long as they are approached in a prudent and balanced manner. An enriched environment is also a contributor to the modulation of hippocampal plasticity, as it enhances cerebral processes through positive pressure to adapt to multiple and diverse environmental stimuli, thus, enabling learning and memory functions.

Nevertheless, the presented non-pharmacological interventions are not intended to represent an exhaustive list of instruments with hippocampal plasticity modulation effects. Studies and findings in this area are ongoing and continuously reveal new ways and means of approaching this sensitive and important matter. At the same time, it has been observed that these interventions have better results when combined in the same protocol. For example, physical exercise alone has beneficial results, but physical exercise in an enriched environment is even more effective in promoting hippocampal plasticity [120] as well as caloric restriction-induced by intermittent fasting merged with physical exercise [116].

6. Conclusions

It is well-established that the optimal functioning and performance of the nervous system and cognitive processes depend on the overall well-being of the body. Imbalanced dietary habits, a sedentary lifestyle, lack of quality sleep, and various environmental conditions and stimuli have been shown in numerous studies on both humans and animals to negatively impact learning and memory. Additionally, the use of prescription drugs for other medical issues or the use of harmful substances like illicit drugs can also have detrimental effects. The hippocampus, a cerebral region known to be associated with these processes, is characterized by its plasticity. While negative factors can negatively impact its performance, the activity of the hippocampus can also be proved or enhanced by exposure to interventions described in this chapter.

Currently, research on the hippocampus is focused on understanding the underlying mechanisms of various brain disorders, such as Alzheimer's disease, temporal lobe epilepsy, and schizophrenia, as well as the neural basis of memory and emotion. One exciting avenue for future research is the study of the neural circuits connecting the hippocampus and the prefrontal cortex, as this can provide insight into how memories are stored and consolidated. Additionally, research on the role of sleep and SWA in the hippocampus is also gaining traction, as it has been shown to play a critical role in memory formation and consolidation. Another promising area of research is the use of neuroimaging techniques, such as functional magnetic resonance imaging (fMRI), to investigate the dynamics of the hippocampus and its subregions in real-time, which could provide a deeper understanding of the neural processes involved in brain disorders and memory formation. Overall, the hippocampus is a complex and multifaceted structure of the brain, and there is still much to be discovered about its functions and the mechanisms that underlie them.

Conflict of interest

The authors declare no conflict of interest.

Author details

Livia Livinț Popa[1,2], Diana Chira[2*], Aida Asaftei[1,2], Stefan Strilciuc[1,2] and Dafin Mureșanu[1,2]

1 Department of Neuroscience, "Iuliu Hațieganu" University of Medicine and Pharmacy, Cluj-Napoca, Romania

2 RoNeuro Institute for Neurological Research and Diagnostic, Cluj-Napoca, Romania

*Address all correspondence to: diana.chira@brainscience.ro

References

[1] Leuner B, Gould E. Structural plasticity and hippocampal function. Annuals Review of Psychology. 2010;**61**:111-140

[2] Horner AJ, Doeller CF. Plasticity of hippocampal memories in humans. Current Opinion in Neurobiology. 2017;**43**:102-109

[3] Couillard-Despres S, Iglseder B, Aigner L. Neurogenesis, cellular plasticity and cognition: The impact of stem cells in the adult and aging brain – A Mini-review. Gerontology. 2011;**57**(6):559-564

[4] Mainardi M, Fusco S, Grassi C. Modulation of hippocampal neural plasticity by glucose-related signaling. Neural Plasticity. 2015;**2015**:657928

[5] Cooper C, Moon HY, van Praag H. On the run for hippocampal plasticity. Cold Spring Harbor Perspectives in Medicine. 2018;**8**(4):a029736

[6] Weerasinghe-Mudiyanselage PDE, Ang MJ, Kang S, Kim JS, Moon C. Structural plasticity of the Hippocampus in neurodegenerative diseases. IJMS. 2022;**23**(6):3349

[7] Zarif H, Nicolas S, Petit-Paitel A, Chabry J, Guyon A. How does an enriched environment impact hippocampus brain plasticity? În: Stuchlik A, editor. The Hippocampus - Plasticity and Functions. London, UK, InTech; 2018. Available from: http://www.intechopen.com/books/the-hippocampus-plasticity-and-functions/how-does-an-enriched-environment-impact-hippocampus-brain-plasticity

[8] Li A, Yau S, Machado S, tWang P, Yuan TF, So KF. Enhancement of hippocampal plasticity by physical exercise as a Polypill for stress and depression: A review. CNSNDDT. 2019;**18**(4):294-306

[9] Wang F, Pan F, Shapiro LA, Huang JH. Stress induced neuroplasticity and mental disorders. Neural Plasticity. 2017;**2017**:1-3

[10] Pittenger C, Duman RS. Stress, depression, and neuroplasticity: A convergence of mechanisms. Neuropsychopharmacology. 2008;**33**(1):88-109

[11] Hayley S, Litteljohn D. Neuroplasticity and the next wave of antidepressant strategies. Frontier in Cell Neuroscience. 2013;**2013**:7. Available from: http://journal.frontiersin.org/article/10.3389/fncel.2013.00218/abstract

[12] Pilar-Cuéllar F, Vidal R, Díaz A, Castro E, dos Anjos S, Pascual-Brazo J, et al. Neural plasticity and proliferation in the generation of antidepressant effects: Hippocampal implication. Neural Plasticity. 2013;**2013**:1-21

[13] Klimesch W. EEG alpha and theta oscillations reflect cognitive and memory performance: A review and analysis. Brain Research Reviews. 1999;**29**(2-3):169-195

[14] Fuchsberger T, Paulsen O. Modulation of hippocampal plasticity in learning and memory. Current Opinion in Neurobiology. 2022;**75**:102558

[15] Huber R, Born J. Sleep, synaptic connectivity, and hippocampal memory during early development. Trends in Cognitive Sciences. 2014;**18**(3):141-152

[16] Halbower AC, Mark ME. Neuropsychological morbidity linked

to childhood sleep-disordered breathing. Sleep Medicine Reviews. 2006;**10**(2):97-107

[17] Schulz KM, Molenda-Figueira HA, Sisk CL. Back to the future: The organizational-activational hypothesis adapted to puberty and adolescence. Hormonal Behaviour. 2009;**55**(5):597-604

[18] Friston K. Functional and effective connectivity in neuroimaging: A synthesis. Human Brain Mapping—Wiley Online Library. 1994;**1994**. Available from: https://onlinelibrary.wiley.com/doi/10.1002/hbm.460020107

[19] Riley JD, Chen EE, Winsell J, Davis EP, Glynn LM, Baram TZ, et al. Network specialization during adolescence: Hippocampal effective connectivity in boys and girls. NeuroImage. 2018;**15**(175):402-412

[20] Lenroot RK, Gogtay N, Greenstein DK, Wells EM, Wallace GL, Clasen LS, et al. Sexual dimorphism of brain developmental trajectories during childhood and adolescence. NeuroImage. 2007;**36**(4):1065-1073

[21] Perrin JS, Hervé PY, Leonard G, Perron M, Pike GB, Pitiot A, et al. Growth of white matter in the adolescent brain: Role of testosterone and androgen receptor. The Journal of Neuroscience. 2008;**28**(38):9519-9524

[22] Huo H, Zhang R, Seger CA, Feng T, Chen Q. The effect of trait anxiety on risk-taking: Functional coupling between right hippocampus and left insula. Psychophysiology. 2020;**57**(10):e13629

[23] Ernst M, Paulus MP. Neurobiology of decision making: A selective review from a neurocognitive and clinical perspective. Biological Psychiatry. 2005;**58**(8):597-604

[24] Blankenstein NE, Peper JS, Crone EA, van Duijvenvoorde ACK. Neural mechanisms underlying risk and ambiguity attitudes. Journal of Cognitive Neuroscience. 2017;**29**(11):1845-1859

[25] Milad MR, Quirk GJ. Fear extinction as a model for translational neuroscience: Ten years of progress. Annual Review of Psychology. 2012;**63**:129-151

[26] Cheng W, Rolls ET, Ruan H, Feng J. Functional connectivities in the brain that mediate the association between depressive problems and sleep quality. JAMA Psychiatry. 2018;**75**(10):1052-1061

[27] Collin SHP, Milivojevic B, Doeller CF. Memory hierarchies map onto the hippocampal long axis in humans. Nature Neuroscience. 2015;**18**(11):1562-1564

[28] Stark SM, Frithsen A, Stark CEL. Age-related alterations in functional connectivity along the longitudinal axis of the hippocampus and its subfields. Hippocampus. 2021;**31**(1):11-27

[29] Blum S, Habeck C, Steffener J. Functional connectivity of the posterior hippocampus is more dominant as we age. Cognitive Neuroscience. 2014;**5**:3-4

[30] Jack CR, Petersen RC, Xu YC, Waring SC, O'Brien PC, Tangalos EG, et al. Medial temporal atrophy on MRI in normal aging and very mild Alzheimer's disease. Neurology. 1997;**49**(3):786-794

[31] Ciftci E, Karacay R, Caglayan A, Altunay S, Ates N, Altintas MO, et al. Neuroprotective effect of lithium in cold- induced traumatic brain injury in mice. Behavioural Brain Research. 2020;**392**:112719

[32] Son H, Yu IT, Hwang SJ, Kim JS, Lee SH, Lee YS, et al. Lithium enhances long-term potentiation independently of hippocampal neurogenesis in

the rat dentate gyrus. Journal of Neurochemistry. 2003;**85**(4):872-881

[33] Kim JS, Chang MY, Yu IT, Kim JH, Lee SH, Lee YS, et al. Lithium selectively increases neuronal differentiation of hippocampal neural progenitor cells both in vitro and in vivo. Journal of Neurochemistry. 2004;**89**(2):324-336

[34] Chen G. Enhancement of hippocampal neurogenesis by lithium - PubMed [Internet]. 2000. Available from: https://pubmed.ncbi.nlm.nih.gov/10987856/

[35] Hajek T, Kopecek M, Höschl C, Alda M. Smaller hippocampal volumes in patients with bipolar disorder are masked by exposure to lithium: A meta-analysis. Journal of Psychiatry & Neuroscience. 2012;**37**(5):333-343

[36] Markou A, Duka T, Prelevic G. Estrogens and brain function. Hormones International Journal of Endocrinology and Metabolism. 2005;**4**(1):9-17

[37] Fester. Proliferation and apoptosis of hippocampal granule cells require local oestrogen synthesis - Fester - 2006. Journal of Neurochemistry - Wiley Online Library [Internet]. 2006 Available from:: https://onlinelibrary.wiley.com/doi/full/10.1111/j.1471-4159.2006.03809.x

[38] Henderson VW. Estrogen-containing hormone therapy and Alzheimer's disease risk: Understanding discrepant inferences from observational and experimental research. Neuroscience. 2006;**138**(3):1031-1039

[39] Cushman M, Legault C, Barrett-Connor E, Stefanick ML, Kessler C, Judd HL, et al. Effect of postmenopausal hormones on inflammation-sensitive proteins: The postmenopausal Estrogen/progestin interventions (PEPI) study. Circulation. 1999;**100**(7):717-722

[40] Nalivaeva NN, Belyaev ND, Turner AJ. Sodium valproate: An old drug with new roles. Trends in Pharmacological Sciences. 2009;**30**(10):509-514

[41] Manji HK, Moore GJ, Chen G. Clinical and preclinical evidence for the neurotrophic effects of mood stabilizers: Implications for the pathophysiology and treatment of manic-depressive illness. Biological Psychiatry. 2000;**48**(8):740-754

[42] Hao Y, Creson T, Zhang L, Li P, Du F, Yuan P, et al. Mood stabilizer valproate promotes ERK pathway-dependent cortical neuronal growth and neurogenesis. The Journal of Neuroscience. 2004;**24**(29):6590-6599

[43] Muresanu DF, Verisezan Rosu O, Popa LL, Strilciuc S, Chira D, Buzoianu A. Psychopharmacotherapy of pain. In: NeuroPsychopharmacotherapy. Cham: Springer; 2021

[44] Luo L, Tan RX. Fluoxetine inhibits dendrite atrophy of hippocampal neurons by decreasing nitric oxide synthase expression in rat depression model. Acta Pharmacologica Sinica. 2001;**22**(10):865-870

[45] Malberg JE, Duman RS. Cell proliferation in adult hippocampus is decreased by inescapable stress: Reversal by fluoxetine treatment. Neuropsychopharmacology. 2003;**28**(9):1562-1571

[46] Manev R, Uz T, Manev H. Fluoxetine increases the content of neurotrophic protein S100beta in the rat hippocampus. European Journal of Pharmacology. 2001;**420**(2-3):R1-R2

[47] Santarelli L, Saxe M, Gross C, Surget A, Battaglia F, Dulawa S, et al. Requirement of hippocampal neurogenesis

for the Behavioral effects of antidepressants. Science. 2003;**301**(5634):805-809

[48] Jung J, Loy K, Schilling EM, Röther M, Brauner J, Huth T, et al. The antidepressant fluoxetine mobilizes vesicles to the recycling Pool of rat hippocampal synapses during high activity. Molecular Neurobiology. 2013;**2013**:49

[49] Abou-Khalil BW. Antiepileptic Drugs. Continuum (Minneap Minn). Febrary. 2016;**22**(1):132-156

[50] Keppel Hesselink JM. Phenytoin: A step by step insight into its multiple mechanisms of action-80 years of mechanistic studies in neuropharmacology. Journal of Neurology. 2017;**264**(9):2043-2047

[51] Zhang YM, Yang Q, Xu CT, Li KS, Li WQ. Effects of phenytoin on morphology and structure of hippocampal CA3 pyramidal neurons of rats in chronic stress. Acta Pharmacologica Sinica. 2003;**24**(5):403-407

[52] Reagan LP, McKittrick CR, McEwen BS. Corticosterone and phenytoin reduce neuronal nitric oxide synthase messenger RNA expression in rat hippocampus. Neuroscience. 1999;**91**(1):211-219

[53] Dhikav V. Can phenytoin prevent Alzheimer's disease? Medical Hypotheses. 2006;**67**(4):725-728

[54] Bremner JD, Mletzko T, Welter S, Quinn S, Williams C, Brummer M, et al. Effects of phenytoin on memory, cognition and brain structure in post-traumatic stress disorder: A pilot study. Journal of Psychopharmacology March 2005;19(2):159-165.

[55] Mureșanu DF, Livinț Popa L, Chira D, Dăbală V, Hapca E, Vlad I, et al. Role and impact of Cerebrolysin for ischemic stroke care. Journal of Clinical Medicine. 2022;**11**(5):1273

[56] Zhang L, Chopp M, Meier DH, Winter S, Wang L, Szalad A, et al. Sonic hedgehog signaling pathway mediates cerebrolysin-improved neurological function after stroke. Stroke. 2013;**44**(7):1965-1972

[57] Kutlu MG, Gould TJ. Effects of drugs of abuse on hippocampal plasticity and hippocampus-dependent learning and memory: Contributions to development and maintenance of addiction. Learning & Memory. 2016;**23**(10):515-533

[58] Dackis CA, O'Brien CP. Cocaine dependence: A disease of the brain's reward centers. Journal of Substance Abuse Treatment. 2001;**21**(3):111-117

[59] Brody SL, Slovis CM, Wrenn KD. Cocaine-related medical problems: Consecutive series of 233 patients. The American Journal of Medicine. 1990;**88**(4):325-331

[60] Haim DY, Lippmann ML, Goldberg SK, Walkenstein MD. The pulmonary complications of crack cocaine. A comprehensive review. Chest. 1995;**107**(1):233-240

[61] Lange RA, Hillis LD. Cardiovascular complications of cocaine use. The New England Journal of Medicine. 2001;**345**(5):351-358

[62] Thomas MJ, Kalivas PW, Shaham Y. Neuroplasticity in the mesolimbic dopamine system and cocaine addiction. British Journal of Pharmacology. 2008;**154**(2):327-342

[63] Anderson SM, Pierce RC. Cocaine-induced alterations in dopamine receptor signaling: Implications for reinforcement and reinstatement. Pharmacology & Therapeutics. 2005;**106**(3):389-403

[64] Rogers JL, See RE. Selective inactivation of the ventral hippocampus attenuates cue-induced and cocaine-primed reinstatement of drug-seeking in rats. Neurobiology of Learning and Memory. 2007;**87**(4):688-692

[65] Ramirez DR, Bell GH, Lasseter HC, Xie X, Traina SA, Fuchs RA. Dorsal hippocampal regulation of memory reconsolidation processes that facilitate drug context-induced cocaine-seeking behavior in rats. The European Journal of Neuroscience. 2009;**30**(5):901-912

[66] Segal M, Richter-Levin G, Maggio N. Stress-induced dynamic routing of hippocampal connectivity: A hypothesis. Hippocampus. 2010;**20**:1332-1338

[67] Abel T, Lattal KM. Molecular mechanisms of memory acquisition, consolidation and retrieval. Current Opinion in Neurobiology. 2001;**11**(2):180-187

[68] Trifilieff P, Herry C, Vanhoutte P, Caboche J, Desmedt A, Riedel G, et al. Foreground contextual fear memory consolidation requires two independent phases of hippocampal ERK/CREB activation. Learning & Memory. 2006;**13**(3):349-358

[69] Heal DJ, Smith SL, Gosden J, Nutt DJ. Amphetamine, past and present – A pharmacological and clinical perspective. Journal of Psychopharmacology. 2013;**27**(6):479-496

[70] Centers for Disease Control and Prevention. Methamphetamine use and risk for HIV/AIDS [Internet]. 2007. Available from: https://stacks.cdc.gov/view/cdc/11778

[71] Guttmann E, Sargant W. Observations on Benzedrine. British Medical Journal. 1937;**1**(3984): 1013-1015

[72] Tidy H. Benzedrine: Uses and Abuses. Proceedings of the Royal Society of Medicine. 1939;**32**(4):385-398

[73] Soetens E, D'Hooge R, Hueting JE. Amphetamine enhances human-memory consolidation. Neuroscience Letters. 1993;**161**(1):9-12

[74] Zeeuws I, Soetens E. Verbal memory performance improved via an acute administration of D-amphetamine. Human Psychopharmacology. 2007;**22**(5):279-287

[75] Johnson BA, Ait-Daoud N, Wells LT. Effects of isradipine, a dihydropyridine-class calcium channel antagonist, on D-methamphetamine-induced cognitive and physiological changes in humans. Neuropsychopharmacology. 2000;**22**(5):504-512

[76] Silber BY, Croft RJ, Papafotiou K, Stough C. The acute effects of d-amphetamine and methamphetamine on attention and psychomotor performance. Psychopharmacology. 2006;**187**(2):154-169

[77] Halliday R, Naylor H, Brandeis D, Callaway E, Yano L, Herzig K. The effect of D-amphetamine, clonidine, and yohimbine on human information processing. Psychophysiology. 1994;**31**(4):331-337

[78] Wilens TE, Adler LA, Adams J, Sgambati S, Rotrosen J, Sawtelle R, et al. Misuse and diversion of stimulants prescribed for ADHD: A systematic review of the literature. Journal of the American Academy of Child and Adolescent Psychiatry. 2008;**47**(1):21-31

[79] Jayanthi S, Daiwile AP, Cadet JL. Neurotoxicity of methamphetamine: Main effects and mechanisms. Experimental Neurology. 2021;**344**:113795

[80] Cha YM, Jones KH, Kuhn CM, Wilson WA, Swartzwelder HS. Sex differences in the effects of delta9-tetrahydrocannabinol on spatial learning in adolescent and adult rats. Behavioural Pharmacology. 2007;**18**(5-6):563-569

[81] Niyuhire F, Varvel SA, Martin BR, Lichtman AH. Exposure to marijuana smoke impairs memory retrieval in mice. The Journal of Pharmacology and Experimental Therapeutics. 2007;**322**(3):1067-1075

[82] Stern CAJ, Gazarini L, Takahashi RN, Guimarães FS, Bertoglio LJ. On disruption of fear memory by reconsolidation blockade: Evidence from cannabidiol treatment. Neuropsychopharmacology. 2012;**37**(9):2132-2142

[83] Ranganathan M, D'Souza D. The acute effects of cannabinoids on memory in humans: A review. Psychopharmacology. 2006;**188**(4). Available from: https://pubmed.ncbi.nlm.nih.gov/17019571/

[84] Campbell KA, Foster TC, Hampson RE, Deadwyler SA. Effects of delta 9-tetrahydrocannabinol on sensory-evoked discharges of granule cells in the dentate gyrus of behaving rats. The Journal of Pharmacology and Experimental Therapeutics. 1986;**239**(3):941-945

[85] Robinson TE, Kolb B. Morphine alters the structure of neurons in the nucleus accumbens and neocortex of rats. Synapse. 1999;**33**(2):160-162

[86] Ornstein TJ, Iddon JL, Baldacchino AM, Sahakian BJ, London M, Everitt BJ, et al. Profiles of cognitive dysfunction in chronic amphetamine and heroin abusers. Neuropsychopharmacology. 2000;**23**(2):113-126

[87] DeCarolis NA, Eisch AJ. Hippocampal neurogenesis as a target for the treatment of mental illness: A critical evaluation. Neuropharmacology. 2010;**58**(6):884-893

[88] Morón JA, Abul-Husn NS, Rozenfeld R, Dolios G, Wang R, Devi LA. Morphine administration alters the profile of hippocampal postsynaptic density-associated proteins: A proteomics study focusing on endocytic proteins. Molecular & Cellular Proteomics. 2007;**6**(1):29-42

[89] Pu H, Wang X, Zhang J, Ma C, Su Y, Li X, et al. Cerebellar neuronal apoptosis in heroin-addicted rats and its molecular mechanism. Int. Journal of Clinical Experimental Pathology. 2015;**8**(7):8260-8267

[90] Ohayon MM, Carskadon MA, Guilleminault C, Vitiello MV. Meta-analysis of quantitative sleep parameters from childhood to old age in healthy individuals: Developing normative sleep values across the human lifespan. Sleep. 2004;**27**(7):1255-1273

[91] Mander BA, Rao V, Lu B, Saletin JM, Lindquist JR, Ancoli-Israel S, et al. Prefrontal atrophy, disrupted NREM slow waves and impaired hippocampal-dependent memory in aging. Nature Neuroscience. 2013;**16**(3):357-364

[92] Bubu OM, Brannick M, Mortimer J, Umasabor-Bubu O, Sebastião YV, Wen Y, et al. Sleep, cognitive impairment, and Alzheimer's disease: A systematic review and Meta-analysis. Sleep. 2017;**40**(1)

[93] Feld G, Diekelmann S. Frontiers | Sleep smart—optimizing sleep for declarative learning and memory [Internet]. 2023. Available from: https://www.frontiersin.org/articles/10.3389/fpsyg.2015.00622/full

[94] Steriade M, Contreras D, Curró Dossi R, Nuñez A. The slow (< 1 Hz) oscillation in reticular thalamic and thalamocortical neurons: Scenario of sleep rhythm generation in interacting thalamic and neocortical networks. The Journal of Neuroscience. 1993;**13**(8):3284-3299

[95] Yavari F, Jamil A, Mosayebi Samani M, Vidor LP, Nitsche MA. Basic and functional effects of transcranial electrical stimulation (tES)-an introduction. Neuroscience and Biobehavioral Reviews. 2018;**85**:81-92

[96] Rohan JG, Miklasevich MK, McInturf SM, Bechmann NA, Moore RJ, Hatcher-Solis C, et al. Polarity and subfield specific effects of transcranial direct current stimulation on hippocampal plasticity. Neurobiology of Learning and Memory. 2020;**167**:107126

[97] Ladenbauer J, Ladenbauer J, Külzow N, de Boor R, Avramova E, Grittner U, et al. Promoting sleep oscillations and their functional coupling by transcranial stimulation enhances memory consolidation in mild cognitive impairment. The Journal of Neuroscience. 2017;**37**(30):7111-7124

[98] Jung SH, Hatcher-Solis C, Moore R, Bechmann N, Harshman S, Martin J, et al. Noninvasive brain stimulation enhances memory acquisition and is associated with Synaptoneurosome modification in the rat Hippocampus. eNeuro. 2019;**6**(6):11-19

[99] Luber B, Lisanby SH. Enhancement of human cognitive performance using transcranial magnetic stimulation (TMS). NeuroImage. 2014;**85**(Pt 3):961-970

[100] Grimaldi D, Papalambros NA, Zee PC, Malkani RG. Neurostimulation techniques to enhance sleep and improve cognition in aging. Neurobiology of Disease. 2020;**141**:104865

[101] Ma J, Zhang Z, Kang L, Geng D, Wang Y, Wang M, et al. Repetitive transcranial magnetic stimulation (rTMS) influences spatial cognition and modulates hippocampal structural synaptic plasticity in aging mice. Experimental Gerontology. 2014;**58**:256-268

[102] Phipps CJ, Murman DL, Warren DE. Stimulating memory: Reviewing interventions using repetitive transcranial magnetic stimulation to enhance or restore memory abilities. Brain Sciences. 2021;**11**(10):1283

[103] Guo JY, Ragland JD, Carter CS. Memory and cognition in schizophrenia. Molecular Psychiatry. 2019;**24**(5):633-642. DOI:10.1038/s41380-018-0231-1

[104] Zhao J, Li Z, Cong Y, Zhang J, Tan M, Zhang H, et al. Repetitive transcranial magnetic stimulation improves cognitive function of Alzheimer's disease patients. Oncotarget. 2017;**8**(20):33864-33871

[105] Amzica F, Steriade M. Disconnection of intracortical synaptic linkages disrupts synchronization of a slow oscillation. Journal of Neuroscience. 1995;**1995**. Available from: https://www.jneurosci.org/content/15/6/4658

[106] Ngo HVV, Martinetz T, Born J, Mölle M. Auditory closed-loop stimulation of the sleep slow oscillation enhances memory. Neuron. 2013;**78**(3):545-553

[107] Mölle M, Born J. Slow oscillations orchestrating fast oscillations and memory consolidation. Progress in Brain Research. 2011;**193**:93-110

[108] Bradley RL, Jeon JY, Liu FF, Maratos-Flier E. Voluntary exercise

improves insulin sensitivity and adipose tissue inflammation in diet-induced obese mice. American Journal of Physiology. Endocrinology and Metabolism. 2008;**295**(3):E586-E594

[109] World Health Organization. Dementia [Internet]. 2020. Available from: https://www.who.int/news-room/fact-sheets/detail/dementia

[110] Liu IT, Lee WJ, Lin SY, Chang ST, Kao CL, Cheng YY. Therapeutic effects of exercise training on elderly patients with dementia: A randomized controlled trial. Achieves in Physical Medical Rehabilitation. 2020;**101**(5):762-769

[111] Babaei P, Azari H. Frontiers | Exercise training improves memory performance in older adults: A narrative review of evidence and possible mechanisms [Internet]. 2022. Available from: https://www.frontiersin.org/articles/10.3389/fnhum.2021.771553/full

[112] Farioli-Vecchioli S, Tirone F. Control of the cell cycle in adult neurogenesis and its relation with physical exercise. Brain Plastic. 2020;**1**(1):41-54

[113] Ma CL, Ma XT, Wang JJ, Liu H, Chen YF, Yang Y. Physical exercise induces hippocampal neurogenesis and prevents cognitive decline. Behavioural Brain Research. 2017;**15**(317):332-339

[114] Hood DA, Memme JM, Oliveira AN, Triolo M. Maintenance of skeletal muscle mitochondria in health, exercise, and aging. Annual Review of Physiology. 2019;**81**(1):19-41

[115] Attuquayefio T, Stevenson RJ, Oaten MJ, Francis HM. A four-day Western-style dietary intervention causes reductions in hippocampal-dependent learning and memory and interoceptive sensitivity. PLoS One. 2017;**12**(2):e0172645

[116] Lee THY, Yau SY. From obesity to hippocampal neurodegeneration: Pathogenesis and non-pharmacological interventions. International Journal of Molecular Sciences. 2020;**22**(1):201

[117] Iwabu M, Yamauchi T, Okada-Iwabu M, Sato K, Nakagawa T, Funata M, et al. Adiponectin and AdipoR1 regulate PGC-1alpha and mitochondria by Ca(2+) and AMPK/SIRT1. Nature. 2010;**464**(7293):1313-1319

[118] Schmidt S, Haase M, Best L, Groth M, Lindner J, Witte OW, et al. Restoring age-related cognitive decline through environmental enrichment: A transcriptomic approach. Cell. 2022;**11**(23):3864

[119] Ali AAH, von Gall C. Adult neurogenesis under control of the circadian system. Cell. 2022;**11**(5):764

[120] Rostami S, Haghparast A, Fayazmilani R. The downstream effects of forced exercise training and voluntary physical activity in an enriched environment on hippocampal plasticity in preadolescent rats. Brain Research. 2021;**1759**:147373

Chapter 7

Learning and Memory Impairment Induced by Amyloid Beta Peptide and Effects of Thymol on Hippocampal Synaptic Plasticity in Rats Fed a High-Fat Diet That Received Amyloid Beta

Masoumeh Asadbegi (Hamedi), Alireza Komaki, Parsa Amiri, Seyed Asaad Karimi, Parichehreh Yaghmaei, Azadeh Ebrahim-Habibi and Iraj Salehi

Abstract

Thymol is a natural phenolic compound that is present in various plants; the significant antioxidant activities of Thymol may be helpful in preventing the progress of various oxidative stress-related diseases. Recent studies have confirmed that antioxidant-rich foods play a vital role in the disease prevention of neurodegenerative diseases, including Alzheimer's disease (AD). We examined the protective and therapeutic effects of Thymol on the Aβ-induced long-term potentiation (LTP) impairments in rats fed a high-fat diet. LTP is a type of synaptic activity that has been thoroughly studied in the hippocampus and is thought to be the neural correlate of learning and memory. If Thymol is protective against AD-related impairments, then natural therapeutic agents based on the structure of Thymol could be used to protect against oxidative stress-related illnesses, such as AD.

Keywords: learning and memory, hippocampal synaptic plasticity, rat, amyloid beta, long-term potentiation, Thymol, high-fat diet

1. Introduction

Turning to a growing public health issue, Alzheimer's disease (AD) is raising concerns worldwide with an increasing rate of occurrence [1, 2]. As the sixth leading cause of all deaths and the fifth in people over 65, AD is an irreversible neurodegenerative disorder marked by a progressive decline in cognition, function, and behavior [3–6].

The contributing factor to this increased prevalence seems to be the global aging population [7]. The number of deaths resulting from AD in Americans over 65 reached 700,000 in 2017. The estimated cost of healthcare and hospice services for those over 65 inflicted with AD or other forms of dementia is expected to exceed $250 billion [3–8].

Neuronal and synaptic loss, progressive memory loss, personality changes, and dementia are among the pathological manifestations of AD, believed to be primarily caused by the accumulation of extracellular beta-amyloid peptides (Aβ) and the intracellular hyperphosphorylated tau proteins [1, 3, 4].

Aβ aggregation is linked with ROS (reactive oxygen species) generation and metabolism disturbance, leading to synaptic dysfunction and neuronal death due to membrane lipid peroxidation [2, 3, 8].

Environmental and genetic factors are also pathogenic contributors, along with nutrition, inflammation, and oxidative stress; however, the AD pathophysiology is not yet fully comprehended. Although a high-fat diet (HFD) is among the risk factors for AD, few studies have been devoted to the link between nutrition and neurodegeneration. It is noteworthy that growing evidence indicates obesity and HFD are associated with cognitive function reduction among the elderly population [1, 9–16].

The dramatic rise in the prevalence of obesity is a grave danger to global healthcare, considering its association with neurodegenerative diseases [17]. One in three American adults is dealing with obesity. Reduced physical activity and diet changes are the outcomes of the modern lifestyle. The cognitive deterioration mechanisms caused by the interaction of aging and obesity are still unclear [16, 18].

Animal studies have shown a connection between a high-fat diet and higher cytokine levels, resulting in a general inflammatory state, increased Aβ aggregation, and behavior impairment [19–22]. The increased AD-like pathogenesis of obesity and HFD, regardless of sex, highlights long-term memory loss and cognitive impairment due to neuronal damage [11, 19]. Addressing obesity and its inflammatory effects could be an invaluable practical method to control the disease [11, 20, 21, 23].

Several complications, such as AD, Parkinson's disease (PD), and aging, are associated with the elevation of oxidative stress, potentially damaging DNA, lipids, sugars, and proteins within cells [1, 7, 23]. The detrimental effects of oxidative stress in AD pathogenesis are well established [24, 25]. Maintaining a balance between oxidation and anti-oxidation through antioxidants can promote neuroprotection and heal the biological system. Thus, oxidative stress and subsequent inflammation are an underlying mechanism in AD pathogenesis, contributing to the Aβ formation. Obesity can generate systemic micro-inflammation, elevated oxidative stress, and reduced hippocampal neurogenesis and function, leading to cognitive deficit [1, 7, 23, 24, 26, 27].

AD severely affects the hippocampus, entorhinal cortex, amygdala, neocortex, and some subcortical areas, damaging synapses and neurons due to high concentrations of plaques and tangles. The hippocampus is crucial in learning and memory, emotion regulation and response, anxiety, stress, and fear [1, 9].

Long-term potentiation (LTP), a sort of long-lasting synaptic plasticity investigated comprehensively in the hippocampus, is a significant study model for learning and memory [28, 29]. Learning and memory are formed through the constant alteration of synaptic communication in the central nervous system. The dentate granule neurons of the hippocampal dentate gyrus (DG), one of the few regions in the rat brain with post-birth neurogenesis, are responsible for learning and memory formation. With the stimulation reaching the perforant pathway (PP), field excitatory postsynaptic potential (EPSP) is produced in the neuron population [29, 30].

Thyme essential oil obtained from Thymus vulgaris L. contains the main monoterpene phenol Thymol (2-isopropyl-5-methyl phenol). Listed in GRAS (generally recognized as safe) as a non-toxic molecule, the compound has antibacterial, anti-inflammatory, antihyperglycemic, hyperlipidemic, and antioxidant properties, beneficial for glycoprotein metabolism regulation in HFD-induced diabetic mice [1, 31].

With 35 million people suffering from AD, a rapidly growing number of afflicted individuals, and no decisive therapy available, more effective therapeutic targets are needed since the existing AD treatment concentrates on the progression delay [32, 33]. On the one hand, limited understanding of risk factors and potential therapeutic targets for neurodegenerative diseases such as AD and, on the other hand, the absence of an effective remedy for obesity-associated brain dysfunction may bring about serious public health ramifications, further emphasizing the identification of relevant preventive and medicinal strategies [2, 17].

In this chapter, we will summarize the current understanding of the AD pathogenesis associated with obesity, recent findings on the cross-link between the two, and their risk factors, which lead to the aggregation of Aβ peptides, synaptic plasticity impairment, and neuronal death. In addition, the potential neuroprotective effects of Thymol, as an option for AD treatment, in rats given HFD will be discussed. We hope to provide significant insights into the development of novel therapeutics for such memory impairments.

2. AD: a significant health problem worldwide

AD, described first in 1906 and named after Alois Alzheimer, is a progressive neurodegenerative disorder and the most prevalent form of dementia associated with old age. Accounting for 50–60% of dementia cases, AD predominantly affects the elderly population over 65 years old. There is an exponential rise with age in afflicted individuals, ranging from 3.0% in 65- to 74-year-olds to 47.2% in those over 85 [12, 34]. The complications involve progressive memory decline, increased apathy, deterioration of intellectual functions, decreased speech function, gait irregularities, and disorientation [34, 35].

The stereotypical characteristic abnormalities include the loss of neurons and synapses, brain atrophy, neurotic plaques or senile plaques (SPs) created due to extracellular Aβ aggregation, the formation of neurofibrillary tangles (NFTs) within neurons in the hippocampus and cortex caused by tau-protein hyperphosphorylation, impaired energy metabolism, mitochondrial dysfunction, elevated activity of prodeath genes and signaling pathways, chronic oxidative stress, and DNA damage [36–40].

It is well established that Aβ results in neurotoxicity and neuronal death; however, the endogenous defense mechanism activated due to Aβ insult is less studied [41]. Early onset of AD affects the medial temporal lobes, hippocampus, and cholinergic neurons of the basal forebrain. The cognitive symptoms in the central nervous system are believed to arise from the loss of cholinergic function, marking a prominent deficit in AD [38, 40].

The increased age-related mortality rate of AD poses a striking social and economic risk to the ever-growing population of AD patients. Although the exact pathogenesis mechanisms remain unclear, oxidative stress and inflammation are hypothesized as the causal mechanisms of AD pathology [12, 34, 39, 40]. Furthermore, apart from age, which seems to be the primary risk factor, diabetes, stroke, atherosclerosis, obesity, and consumption of a high-fat diet are among the risk factors for AD [12, 13].

2.1 Oxidative stress as a trigger of AD

The late-onset sporadic AD, the most common form, is supposed to rise from a combination of genetic susceptibility factors and environmental triggers, with oxidative damage and a slow inflammatory process accepted widely as possible involved mechanisms [42].

Oxidative stress is the imbalance between ROS production and the antioxidant defense system. It is known that oxidative stress increases in parallel with age, smoking, hyperhomocysteinemia, and insulin resistance, thus causing an insulin action impairment in type 2 diabetics, probably due to membrane fluidity alterations, decreased availability of nitric oxide, and increased intracellular calcium content [43].

Furthermore, oxidative stress is assumed to be a primary risk factor and a trigger for AD pathology. Whether preceding oxidative damage is directly responsible for the accumulation of intracellular beta-amyloid 1–42 (Aβ 1–42) remains unclear. Concerning the studies, mitochondria are among the first affected organelles by oxidative stress and Aβ 1–42 toxicity. These organelles serve as accumulation sites for Aβ 1–42, promoting mitochondrial dysfunction and hindering energy metabolism [44].

2.2 AD and inflammation

Numerous amyloid plaque proteins are involved in an inflammatory response in AD, such as pro-inflammatory cytokines, acute-phase proteins, and activated complement factors, most having pleiotropic effects depending on their concentrations, making it hard to assess their contribution to the amyloid formation. In addition, neurons have been shown to play an active role in the neuroinflammatory process of AD [45].

Damage signals such as infection, trauma, redox iron, oxidative agents, and t and b-amyloid oligomers induce neuroinflammation response. Regarding the relationship between the inflammatory process of the brain and neuronal damage, the overproduction of pro-inflammatory agents is linked with the progressive activation of microglial cells and astrocytes [8].

Neuroinflammation and microglia–neuron cross-talks present promising therapeutic targets for the treatment of AD. Studies have demonstrated that long-term treatment with anti-inflammatory drugs decreases AD progression via the disruption of the inflammatory response and t protein self-aggregation. Also, the reduced incidence of AD in patients subjected to anti-inflammatory prescription supports the neuroinflammatory hypothesis [8].

2.3 AD and cholinergic deficit: cholinergic deficit as a consistent and early finding in AD

The neuropeptide/neurotransmitter systems are significantly affected by metabolic status [46]. The established evidence supports the association of cholinergic deficit with the pathogenesis of AD. Acetylcholinesterase inhibitors (AChEIs) have been used to treat AD symptoms through their acetylcholine-medicated boost of neuron-to-neuron transmission. AChEIs also promote antioxidant production against free radical toxicity and β-amyloid-induced injury and suppress cytokine release from monocytes and microglia through the cholinergic anti-inflammatory pathway [47].

Although the cholinergic hypothesis has garnered substantial support and led to the production of the first licensed medication for AD symptoms, it targets a consequential symptom of AD, not the underlying pathological cause, and amyloid neurotoxicity intensifies in the presence of AChE. Regardless, donepezil, rivastigmine, and galantamine are licensed in the UK [47].

The study of AChEI drugs has shown that cholinergic pathways in the cerebral cortex and basal forebrain are compromised in AD, resulting in cholinergic deficit and cognitive impairment; however, growing evidence suggests more of an anti-inflammatory role for AChEIs through the inhibition of cytokine release from activated microglia, leaving the door open to further research. Multiple cholinesterase inhibitors (ChEI) are being developed, including ChEIs, naturally derived ChEIs, hybrids, and synthetic analogs. AD also affects many other neurotransmitters, with their clinical importance not fully clarified [42, 47–49].

3. Obesity: one of the most severe global health problems

High-fat diet consumption continues to increase human obesity, affecting 2 billion worldwide. Aside from the association with overweight pathology, obesity is considered a risk factor for dementia and neurodegenerative disorders and is associated with cognitive impairment, with some deeming the relation controversial [11, 15, 38, 50].

Few from the field of nutrition have turned their attention to neurodegenerative diseases [39].

Gustafson *et al*. considered obesity at older ages (ages seventy nine to eighty eight) as a risk factor for dementia, notably AD. A high-fat diet consumption reduces hippocampal neurogenesis, impairing attention and visual memory even in the short term. Although the underlying causes are not elucidated, a high-fat diet results in cognitive deficits and reduced hippocampal function [11, 12, 18].

Epidemiologic studies also argued that diets rich in saturated fats (especially in midlife) are a primary risk factor for AD development [51]. HFD-induced obesity may impact synaptic plasticity via insulin resistance and altered glucose metabolism, affecting learning, memory, and neuronal survival [14]. HF diet can also have detrimental effects on the brain independent of the explained factors, such as inducing Aβ deposition in the brains of mice or increased expression of amyloid precursor protein (APP) in humans [22].

A greater prevalence of AD is observed in countries with a higher intake of high-fat or high-calorie diets. Studies on transgenic AD mice on a high-fat diet show heightened disease neuropathology and behavioral deficits [12, 13]. Aside from the increased accumulation of the toxic Aβ, animal models show increased susceptibility to HFD-induced bodyweight gain following AD [19, 52].

Investigations in genetically obese animal models indicated impaired spatial memory and hippocampal synaptic plasticity. Since the consumption of HFD in humans is the common cause of obesity, animal models of HFD-induced obesity imitate the pathological obesity changes better [18].

3.1 HFD and oxidative stress

Oxidative stress is caused either by excessive ROS production or by antioxidant defense deficiency [53]. HFD has been shown to induce oxidative stress, mitochondrial dysfunction, inflammation, and adipokine dyshomeostasis, leading to neurodegeneration [54]. As a worldwide epidemic, obesity is characterized by excessive fat deposition, increased cardiovascular risk factors, and high oxidative stress. The early phases of being overweight involve increased ROS production, reduced NO bioavailability, and endothelial dysfunction. Unlike synthetic antioxidant supplementation, diets enriched in natural antioxidants help regulate blood pressure, serum lipid composition, and oxidative stress [55].

3.2 HFD and inflammation

An essential component of neurodegenerative diseases is neuroinflammation. Neuroinflammation caused by the infiltration of inflammatory immune cells and activation of microglial cells, stress response, and the disruption of the blood–brain barrier (BBB) are probable outcomes of the increased metabolic flux in the brain as a result of obesity [17]. HFD consumption can also contribute to autoimmune encephalomyelitis (EAE), highlighting its significant effects on neuroinflammation further [56].

The increased cross-talk between the peripheral system and neuroinflammation is another consequence of HFD consumption, exhibiting an increased induction of pro-inflammatory cytokines in peripheral tissues and the hypothalamus, including interleukin (IL-1b), IL6, and tumor necrosis factor a (TNFa). Investigations show a connection between AD and the impairment of cognition and memory with HFD-induced neuroinflammation. The mechanisms through which obesity promotes neuroinflammation require further investigations [16, 17].

3.3 HFD impairs learning, memory, and synaptic plasticity

Unlike the hypothalamus, few studies investigated the effects of HFD on the hippocampus. Several articles have shown the impacts of obesity and HFD consumption on increased memory impairment, AD, and dementia [2]. The activation of NADPH oxidase activity caused by carbohydrate-enriched HFD consumption was shown to impair learning, memory, and synaptic plasticity [11]. Neurodegeneration is the neuronal loss of structure, function, and death within the regions associated with learning, memory, and emotion, such as the hippocampus and basal forebrain. The progressive nature of neurodegenerative diseases will eventually result in short-term memory loss, mood changes, and cognitive impairment. Memory formation depends on long-term potentiation (LTP), a form of synaptic plasticity. AD patients also suffer from LTP impairment [2]. AD development and reduction of spatial learning skills and hippocampal plasticity have been linked to HFD-induced obesity in rodents, with the promoted brain dysfunction considered a consequence of stress and neuroinflammation [17, 39]. Systemic inflammation and oxidative stress in obesity worsen with age, disrupting the BBB even further. In another study, the neuroinflammation and oxidative stress in aged obese animals contributed to a significant cognitive decline and learning and memory impairment [11, 16].

4. Oxidative stress and AD

4.1 The production of free radicals

A free radical is a highly reactive molecule with an unpaired electron. The free radical binds to another molecule to acquire an electron, thus changing or breaking it biochemically into another free radical or an altered chemical structure. A free radical can damage any molecule, including proteins, carbohydrates, lipids, and nucleic acids, resulting in apoptosis and cell death [26, 57].

Oxidative stress is responsible for the pathogenesis of several diseases, including memory deficits caused by oxidative damage in rats and humans. The brain undergoes severe damage under oxidative stress due to the abundance of unsaturated fatty acids, lower antioxidant protection, redox-active metals (Fe, Cu), and high oxygen concentrations [7, 23, 29].

Aging contributes to increased oxidative stress, probably playing a crucial role in the pathogenesis of AD [7]. The neuron capacity to regulate redox imbalance decreases with age, which might lead to irreversible impairment such as neurodegenerative diseases. MDA, HNE, carbonyls, and other indicators of oxidative stress increase in aging and neurodegenerative diseases; however, a quantifiable method to identify relevant complications is not developed [26].

Studies have recently emphasized the crucial role of oxidative stress in the pathogenesis of various neurodegenerative diseases, such as AD, PD, and amyotrophic lateral sclerosis. Several common indicators are observed, including glutathione loss, DNA, and protein damage [7].

4.2 Antioxidant systems

Antioxidants have shown a promising effect on Aβ-induced neurotoxicity and cell death, improving impaired cognition and memory in AD. Antioxidants protect cells against damage induced by free radicals, preventing neuronal injury due to oxidative stress. The oxidative injury seems to be the factor initiating the neurodegeneration, not merely a byproduct or an end product of the process. However, AD treatment via antioxidants is met with great suspicion, as most of them can hardly cross the BBB. Accordingly, smaller antioxidant molecules or non-toxic carriers can provide favorable modifications for the current complications [25, 26, 29].

ROS substances, antioxidants, and non-enzymatic molecules can be used to assess oxidative stress status. The accepted evaluation method involves the total antioxidant status (TAS) and total oxidant status (TOS) measurements [58].

5. Hippocampus synaptic plasticity and LTP

Hippocampal synaptic plasticity is the principal means of information processing and memory formation [59]. The hippocampus goes through synaptic reorganization and neurophysiological changes in response to stimuli [60]. In the hippocampal input region, the DG is crucial to memory formation, converting the cortical inputs to new output, which then travels to the CA3 [61, 62].

Neurogenesis is preserved in various parts of the adult brain, especially the subventricular zone of the lateral ventricles and the subgranular zone of the DG. The process is induced under different physiological or pathological circumstances, ranging from exercise and environmental adaptations to seizures or injuries [63].

LTP is a continuous increase in synaptic strength, mainly through stimulation. Analyzing LTP in the hippocampus is a widely used cellular and molecular method to evaluate learning and memory. Coordinated gene transcription, protein synthesis, and degradation are needed for long-lasting synaptic plasticity [7, 64]. LTP induction relies on several mechanisms, such as the activation of AMPA and NMDA receptors, changes in the number and shape of the spines, and the enhancement of transmitter release [30, 64].

5.1 Synaptic plasticity and AD

Aβ1–42 has a central role in the development of AD and its cognitive impairment, considerably inhibiting the hippocampal LTP at a cellular level [7, 65]. Reduced LTP and enhanced long-term depression (LTD) have been shown in acute exposure to Aβ, keeping the basal synaptic transmission unchanged. Studies suggest a shared molecular pathway between apoptosis and non-apoptotic functions, including synaptic plasticity. The activation of caspase-3 is essential for LTD and Aβ-induced inhibition of LTP [9, 66].

5.2 Synaptic plasticity and HFD

HFD-induced obesity has several detrimental effects on the hippocampal structure and function, affecting the normal growth of the CNS; impairing LTP, cognitive function, and learning; and disrupting neurogenesis [62, 67].

6. Thymol and potential to treat AD

Thymol obtained from thyme essential oil is a translucent crystal and a monoterpene aromatic alcohol with a boiling point of 232°C. It also has an herbaceous, sweet-medicinal, warm odor and a pleasant taste. The vast biological properties of Thymol include larvicidal, nematicidal, acaricidal, antifungal, antibacterial, anti-inflammatory, and antioxidant activities [68, 69].

6.1 Thymol and antioxidant activity

In the aerobic environment, reactive oxygen species are the most lethal by-products of metabolism that mediate many human diseases including AD, Parkinson's disease, diabetes mellitus, atherosclerosis, and aging processes [70, 71].

Antioxidants protect cell membranes against free radicals and accelerate the excretion of cellular wastes [69]. An oxidant and antioxidant balance is crucial for optimal physiological conditions. Therefore, oxidative stress can be described as redox signaling and control disruption. Such a devised definition can lead to new treatments for oxidative stress-related diseases. Several pieces of research have been devoted to replacing synthetic chemicals with natural substances, like natural antioxidants. Due to significant phenolic content, Thymol has a remarkable reducing power and great ferric-reducing ability. It also shows superb antioxidant properties through extensive scavenging activity. Its total amount of phenol is 0.36 ± 0.06 μg/ml. As a folkloric medicine, it might be helpful to prevent the progression of various diseases related to oxidative stress [24, 69–71].

DPPH radical scavenging activity was used to determine the antioxidant activity. The method represents a strong absorption maximum based on the reduction of DPPH at 517 nm. A hydrogen donor acts as a free radical scavenging antioxidant to pair with the odd electron and decrease absorption strength. Considering the number of captured electrons, Thymol presence shows a stochiometric de-colorization. The reducing properties are demonstrated to be generally associated with the reductones, exerting antioxidant action by breaking the FR chain through the donation of a hydrogen atom. FRAP assay was also used to evaluate the ability of phytochemicals to reduce ferric ions. Thymol displayed a good reducing ability of ferric tripyridyl triazine complex into ferrous-(TPTZ) complex.

The results of the experiments approved a good antioxidant power and a free radical scavenging activity for Thymol. Hence, it can be a basis for herbal medicine development to prevent and treat disorders related to oxidative stress, such as Alzheimer's disease [70].

6.2 Thymol and anti-inflammatory activity

As a natural monoterpene, it has considerable biological influences on cells, mainly through its antioxidant and anti-inflammatory effects. Airway inflammation in ovalbumin-induced mouse asthma was ameliorated by inhibition of NF-κB activation. Via the interference with the activation of nuclear factor kappa (NF-κB) and mitogen-activated protein kinase (MAPK) signaling pathways, the compound exerted its anti-inflammatory properties in lipopolysaccharide-stimulated mouse mammary epithelial cells. The results indicate the anti-inflammatory effects of Thymol through the suppression of several biomarkers involved in inflammation. Thymol is suggested to have significantly ameliorated inflammatory responses through protective anti-oxidation, anti-inflammation, and anti-lipid peroxidation effects. Therefore, it is a promising compound to cure inflammatory processes [72].

6.3 Thymol and anticholinesterase activity

The inhibition of AChE, the predominant enzyme involved in the hydrolysis of acetylcholine (ACh), is a developed therapeutic strategy for AD treatment. Regarding traditional medicine, several plants are reputed to enhance cognitive function and alleviate other symptoms of AD, including depression. T. vulgaris essential oil indicates neuroprotective effects. With a small molecular size and lipophilicity, volatile constituents of essential oils and aglycones from glycosides are likely to cross the BBB. Thymol, carvacrol, and their derivatives, such as thymoquinone and thymohydroquinone, can be used as inhibitors of AChE, though their possible application in Alzheimer's treatment or other cognitive disorders needs further investigation. The probable link between the antioxidant and the AChE inhibitory activity of the mentioned compounds could also be interesting to study [73].

6.4 Thymol and anti-obesity activity

Thymol anti-hyperglycemic and anti-hyperlipidemic activities are yet to be explored [74]. As mentioned, obesity has become a worldwide health problem, and most synthetic anti-obesity drugs have failed to address the issue due to their ineffectiveness or adverse effects. Thymol prevents HFD-induced obesity in the murine model through several mechanisms, including the attenuation of visceral fat accumulation, lipid-lowering action, improvement of insulin and leptin sensitivity, and enhanced antioxidant potential [75].

According to a study, rats given HFD exhibited significant enhancement of body weight gain ($p < 0.001$), visceral pad weight, lipids, alanine aminotransferase (ALT), aspartate amino transaminase (AST), lactate dehydrogenase (LDH), blood urea nitrogen (BUN), glucose, insulin, and leptin levels compared to rats given a standard diet. Thymol treatment showed a significant decrease in ($p < 0.001$) body weight gain, visceral fat-pad weights, lipids, ALT, AST, LDH, BUN, glucose, insulin, and leptin levels in HFD-induced obese rats.

Furthermore, the treatment significantly decreased serum lipid peroxidation and increased antioxidant levels in HFD-induced obese rats, preventing HFD-induced obesity in the murine model through the discussed mechanisms. Interestingly, Thymol may also exhibit promising anti-diabetic effects [74, 75].

Recent studies have provided preliminary positive evidence for the effectiveness and safety of Thymol in alleviating cognitive impairments caused by increased Aβ levels or cholinergic hypofunction [76]. As a whole, it seems that antioxidant, anti-inflammatory, and anticholinesterase activities of Thymol might contribute to its beneficial effects. Our findings suggest that Thymol may be potentially a valuable source of natural therapeutic agents for AD treatment. However, further investigations are necessary to establish its efficacy and potential toxicity in clinical trials [76–78].

7. Conclusions

Despite the ever-increasing studies on AD, no final remedy has been developed, with palliative treatments presenting the best treatment options against AD symptoms until now. Considering the worldwide growth in obesity prevalence and the increasing number of dementia cases in an aging population, such a dual model may exhibit better results than the classical AD models.

In this case, Thymol presented a promising therapeutic potential against AD and HFD consumption that remains to be further investigated. It was hypothesized that antioxidant, anti-inflammatory, anticholinesterase, antihyperglycemic, and hyperlipidemic activities of Thymol may have contributed to its beneficial effects on learning and memory impairment, synaptic plasticity, oxidative stress, and tissue changes in HFD-induced animal models of AD.

The results stress the benefits of Thymol as a dietary antioxidant, providing preliminary evidence for its effectiveness as a remedy against LTP impairments caused by the Aβ aggregation in AD rats given HFD. The antioxidant activity of Thymol may be the reason behind its beneficial effects on hippocampal synaptic plasticity. Therefore, natural therapeutic agents based on Thymol could be used to prevent and cure complications related to oxidative stress, such as AD. Further research is required to establish its efficacy and potential toxicity in clinical trials.

Acknowledgements

My lovely Mehrsa, Majid, and Mom, thank you for all your love, patience, support, and guidance. I would not have concentrated and written this chapter through recent very challenging months without any of you.

I am forever grateful.

Conflict of interest

The authors declare no conflict of interest.

DOI: http://dx.doi.org/10.5772/intechopen.110416

Acronyms and abbreviations

Aβ	amyloid beta
AChE	acetylcholinesterase
AChEIs	acetylcholinesterase inhibitors
AD	Alzheimer's disease
ALT	alanine aminotransferase
APP	amyloid precursor protein
AST	aspartate amino transaminase
BBB	blood–brain barrier
BUN	blood urea nitrogen
CNS	central nervous system
DG	dentate gyrus
EAE	experimental autoimmune encephalomyelitis
EPSP	excitatory postsynaptic potential
GRAS	generally recognized as safe
HFD	high-fat diet
IL	interleukin
LDH	lactate dehydrogenase
LDL	low-density lipoprotein cholesterol
LTD	long-term depression
LTP	long-term potentiation
MAPK	mitogen-activated protein kinase
MCI	mild cognitive impairment
NF-κB	nuclear factor kappa
NFTs	neurofibrillary tangles
NMDA	N-methyl-D-aspartate
Nrf2	nuclear factor
PD	Parkinson's disease
PP	perforant pathway
ROS	reactive oxygen species
SPs	senile plaques
TAS	total antioxidant status
TOS	total oxidant status
TNFa	tumor necrosis factor a

Author details

Masoumeh Asadbegi (Hamedi)[1*], Alireza Komaki[1], Parsa Amiri[1], Seyed Asaad Karimi[1,2,3], Parichehreh Yaghmaei[4], Azadeh Ebrahim-Habibi[4,5,6,7] and Iraj Salehi[1]

1 Neurophysiology Research Center, Hamadan University of Medical Sciences, Hamadan, Iran

2 Department of Physiology, School of Medicine, Shahid Beheshti University of Medical Sciences, Tehran, Iran

3 Neurophysiology Research Center, School of Medicine, Shahid Beheshti University of Medical Sciences, Tehran, Iran

4 Department of Biology, Science and Research Branch, Islamic Azad University, Tehran, Iran

5 Biosensor Research Center, Endocrinology and Metabolism Molecular-Cellular Sciences Institute, Tehran University of Medical Sciences, Tehran, Iran

6 Endocrinology and Metabolism Research Center, Endocrinology and Metabolism Clinical Sciences Institute, Tehran University of Medical Sciences, Tehran, Iran

7 Department of Neuroscience, School of Science and Advanced Technologies in Medicine, Hamadan University of Medical Sciences, Hamadan, Iran

*Address all correspondence to: masoumehasadbegi@gmail.com

References

[1] Asadbegi M, Yaghmaei P, Salehi I, Komaki A, Ebrahim-Habibi A. Investigation of thymol effect on learning and memory impairment induced by intrahippocampal injection of amyloid beta peptide in high fat diet-fed rats. Metabolic Brain Disease. 2017;**32**(3):827-839

[2] Millette D. Short-Term Feeding Effects of High-Fat Diet on Hippocampal Neuroinflammation: Differences among C57BL/6J, C3H/HeN, and C3H/HeJ Mice. Chapel Hill, NC, United States: The University of North Carolina at Chapel Hill; 2012

[3] Ganai SA, Banday S, Farooq Z, Altaf M. Modulating epigenetic HAT activity for reinstating acetylation homeostasis: A promising therapeutic strategy for neurological disorders. Pharmacology & Therapeutics. 2016;**166**:106-122

[4] Huang C, Dong D, Jiao Q, Pan H, Ma L, Wang R. Sarsasapogenin-AA 13 ameliorates Aβ-induced cognitive deficits via improving neuroglial capacity on Aβ clearance and antiinflammation. CNS Neuroscience & Therapeutics. 2017;**23**(6):498-509

[5] Majid T, Griffin D, Criss Z II, Jarpe M, Pautler RG. Pharmocologic treatment with histone deacetylase 6 inhibitor (ACY-738) recovers Alzheimer's disease phenotype in amyloid precursor protein/presenilin 1 (APP/PS1) mice. Alzheimer's & Dementia: Translational Research & Clinical Interventions. 2015;**1**(3):170-181

[6] Reitz C. Alzheimer's disease and the amyloid cascade hypothesis: A critical review. International Journal of Alzheimer's Disease. 17 Mar 2012;**2012**:11. Article ID: 369808. DOI: 10.1155/2012/369808

[7] Komaki H, Faraji N, Komaki A, Shahidi S, Etaee F, Raoufi S, et al. Investigation of protective effects of coenzyme Q10 on impaired synaptic plasticity in a male rat model of Alzheimer's disease. Brain Research Bulletin. 2019;**147**:14-21

[8] Morales I, Guzmán-Martínez L, Cerda-Troncoso C, Farías GA, Maccioni RB. Neuroinflammation in the pathogenesis of Alzheimer's disease. A rational framework for the search of novel therapeutic approaches. Frontiers in Cellular Neuroscience. 2014;**8**:112

[9] Revett TJ, Baker GB, Jhamandas J, Kar S. Glutamate system, amyloid β peptides and tau protein: Functional interrelationships and relevance to Alzheimer disease pathology. Journal of Psychiatry & Neuroscience. 2013;**38**(1):6-23

[10] Mirzaei F, Khazaei M, Komaki A, Amiri I, Jalili C. Virgin coconut oil (VCO) by normalizing NLRP3 inflammasome showed potential neuroprotective effects in amyloid-β induced toxicity and high-fat diet fed rat. Food and Chemical Toxicology. 2018;**118**:68-83

[11] Kim HG, Jeong HU, Park G, Kim H, Lim Y, Oh MS. Mori folium and mori fructus mixture attenuates high-fat diet-induced cognitive deficits in mice. Evidence-based Complementary and Alternative Medicine. 2015;**2015**:8. Article ID: 379418. DOI: 10.1155/2015/379418

[12] Knight EM, Martins IV, Gümüsgöz S, Allan SM, Lawrence CB. High-fat diet-induced memory impairment in triple-transgenic Alzheimer's disease (3xTgAD) mice is independent of changes in amyloid and tau pathology. Neurobiology of Aging. 2014;**35**(8):1821-1832

[13] Martin SA, Jameson CH, Allan SM, Lawrence CB. Maternal high-fat diet worsens memory deficits in the triple-transgenic (3xTgAD) mouse model of Alzheimer's disease. PLoS One. 2014;**9**(6):e99226

[14] McNeilly A, Williamson R, Balfour D, Stewart C, Sutherland C. A high-fat-diet-induced cognitive deficit in rats that is not prevented by improving insulin sensitivity with metformin. Diabetologia. 2012;**55**(11):3061-3070

[15] Moy GA, McNay EC. Caffeine prevents weight gain and cognitive impairment caused by a high-fat diet while elevating hippocampal BDNF. Physiology & Behavior. 2013;**109**:69-74

[16] Tucsek Z, Toth P, Sosnowska D, Gautam T, Mitschelen M, Koller A, et al. Obesity in aging exacerbates blood–brain barrier disruption, neuroinflammation, and oxidative stress in the mouse hippocampus: Effects on expression of genes involved in beta-amyloid generation and Alzheimer's disease. The Journals of Gerontology. Series A, Biological Sciences and Medical Sciences. 2014;**69**(10):1212-1226

[17] Nerurkar PV, Johns LM, Buesa LM, Kipyakwai G, Volper E, Sato R, et al. Momordica charantia (bitter melon) attenuates high-fat diet-associated oxidative stress and neuroinflammation. Journal of Neuroinflammation. 2011;**8**(1):1-19

[18] Hwang LL, Wang CH, Li TL, Chang SD, Lin LC, Chen CP, et al. Sex differences in high-fat diet-induced obesity, metabolic alterations and learning, and synaptic plasticity deficits in mice. Obesity. 2010;**18**(3):463-469

[19] Barron AM, Rosario ER, Elteriefi R, Pike CJ. Sex-specific effects of high fat diet on indices of metabolic syndrome in 3xTg-AD mice: Implications for Alzheimer's disease. PLoS One. 2013;**8**(10):e78554

[20] Lotfi P, Yaghmaei P, Ebrahim-Habibi A. Cymene and metformin treatment effect on biochemical parameters of male NMRI mice fed with high fat diet. Journal of Diabetes and Metabolic Disorders. 2015;**14**(1):1-5

[21] Miller AA, Spencer SJ. Obesity and neuroinflammation: A pathway to cognitive impairment. Brain, Behavior, and Immunity. 2014;**42**:10-21

[22] Vandal M, White PJ, Tremblay C, St-Amour I, Chevrier G, Emond V, et al. Insulin reverses the high-fat diet–induced increase in brain Aβ and improves memory in an animal model of Alzheimer disease. Diabetes. 2014;**63**(12):4291-4301

[23] Asadbegi M, Komaki A, Salehi I, Yaghmaei P, Ebrahim-Habibi A, Shahidi S, et al. Effects of thymol on amyloid-β-induced impairments in hippocampal synaptic plasticity in rats fed a high-fat diet. Brain Research Bulletin. 2018;**137**:338-350

[24] Jones DP. Redefining oxidative stress. Antioxidants & Redox Signaling. 2006;**8**(9-10):1865-1879

[25] Wang H, Wang H, Cheng H, Che Z. Ameliorating effect of luteolin on memory impairment in an Alzheimer's disease model. Molecular Medicine Reports. 2016;**13**(5):4215-4220

[26] Aliev G, Obrenovich ME, Reddy VP, Shenk JC, Moreira PI, Nunomura A, et al. Antioxidant therapy in Alzheimer's disease: Theory and practice. Mini Reviews in Medicinal Chemistry. 2008;**8**(13):1395-1406

[27] Zargooshnia S, Shahidi S, Ghahremanitamadon F, Nikkhah A,

Mehdizadeh M, Soleimani AS. The protective effect of Borago Officinalis extract on amyloid β (25-35)-induced long term potentiation disruption in the dentate gyrus of male rats. Metabolic Brain Disease. 2015;**30**(1):151-156

[28] Bartsch T, Wulff P. The hippocampus in aging and disease: From plasticity to vulnerability. Neuroscience. 19 Nov 2015;**309**:1-16

[29] Omidi G, Rezvani-Kamran A, Ganji A, Komaki S, Etaee F, Asadbegi M, et al. Effects of Hypericum scabrum extract on dentate gyrus synaptic plasticity in high fat diet-fed rats. The Journal of Physiological Sciences. 2020;**70**(1):1-8

[30] Komaki A, Khalili A, Salehi I, Shahidi S, Sarihi A. Effects of exposure to an extremely low frequency electromagnetic field on hippocampal long-term potentiation in rat. Brain Research. 2014;**1564**:1-8

[31] Villanueva Bermejo D, Angelov I, Vicente G, Stateva RP, Rodriguez García-Risco M, Reglero G, et al. Extraction of thymol from different varieties of thyme plants using green solvents. Journal of the Science of Food and Agriculture. 2015;**95**(14):2901-2907

[32] Etaee F, Asadbegi M, Taslimi Z, Shahidi S, Sarihi A, Asl SS, et al. The effects of methamphetamine and buprenorphine, and their interaction on anxiety-like behavior and locomotion in male rats. Neuroscience Letters. 2017;**655**:172-178

[33] Reinke AA, Gestwicki JE. Insight into amyloid structure using chemical probes. Chemical Biology & Drug Design. 2011;77(6):399-411

[34] Singhal AK, Naithani V, Bangar OP. Medicinal plants with a potential to treat Alzheimer and associated symptoms. International Journal of Nutrition, Pharmacology, Neurological Diseases. 2012;**2**(2):84

[35] Herculano B, Tamura M, Ohba A, Shimatani M, Kutsuna N, Hisatsune T. β-Alanyl-L-histidine rescues cognitive deficits caused by feeding a high fat diet in a transgenic mouse model of Alzheimer's disease. Journal of Alzheimer's Disease. 2013;**33**(4):983-997

[36] Cai Z, Yan Y, Wang Y. Minocycline alleviates beta-amyloid protein and tau pathology via restraining neuroinflammation induced by diabetic metabolic disorder. Clinical Interventions in Aging. 2013;**8**:1089

[37] De la Monte SM, Wands JR. Alzheimer's disease is type 3 diabetes—Evidence reviewed. Journal of Diabetes Science and Technology. 2008;**2**(6):1101-1113

[38] Henderson VW. Alzheimer's disease: Review of hormone therapy trials and implications for treatment and prevention after menopause. The Journal of steroid biochemistry and molecular biology. 2014;**142**:99-106

[39] Morris MC, Tangney CC. Dietary fat composition and dementia risk. Neurobiology of Aging. 2014;**35**:S59-S64

[40] Peng D, Pan X, Cui J, Ren Y, Zhang J. Hyperphosphorylation of tau protein in hippocampus of central insulin-resistant rats is associated with cognitive impairment. Cellular Physiology and Biochemistry. 2013;**32**(5):1417-1425

[41] Tao CC, Hsu WL, Ma YL, Cheng SJ, Lee EH. Epigenetic regulation of HDAC1 SUMOylation as an endogenous neuroprotection against Aβ toxicity in a mouse model of Alzheimer's disease. Cell Death and Differentiation. 2017;**24**(4):597-614

[42] McGleenon B, Dynan K, Passmore A. Acetylcholinesterase inhibitors in Alzheimer's disease. British Journal of Clinical Pharmacology. 1999;**48**(4):471

[43] Yilmaz M, Bukan N, Gk A, Karakoç A, Toruner F, Çakir N, et al. The effects of rosiglitazone and metformin on oxidative stress and homocysteine levels in lean patients with polycystic ovary syndrome. Human Reproduction. 2005;**20**(12):3333-3340

[44] Hernández-Zimbrón L, Rivas-Arancibia S. Oxidative stress caused by ozone exposure induces β-amyloid 1-42 overproduction and mitochondrial accumulation by activating the amyloidogenic pathway. Neuroscience. 2015;**304**:340-348

[45] Eikelenboom P, Bate C, Van Gool W, Hoozemans J, Rozemuller J, Veerhuis R, et al. Neuroinflammation in Alzheimer's disease and prion disease. Glia. 2002;**40**(2):232-239

[46] Gahete MD, Córdoba-Chacón J, Kineman RD, Luque RM, Castaño JP. Role of ghrelin system in neuroprotection and cognitive functions: Implications in Alzheimer's disease. Peptides. 2011;**32**(11):2225-2228

[47] Tabet N. Acetylcholinesterase inhibitors for Alzheimer's disease: Anti-inflammatories in acetylcholine clothing! Age and Ageing. 2006;**35**(4):336-338

[48] Talesa VN. Acetylcholinesterase in Alzheimer's disease. Mechanisms of Ageing and Development. 2001;**122**(16):1961-1969

[49] Mehta M, Adem A, Sabbagh M. New acetylcholinesterase inhibitors for Alzheimer's disease. International Journal of Alzheimer's Disease. 2012;**2012**:8. Article ID: 728983. DOI: 10.1155/2012/728983

[50] Petrov D, Pedrós I, Artiach G, Sureda FX, Barroso E, Pallàs M, et al. High-fat diet-induced deregulation of hippocampal insulin signaling and mitochondrial homeostasis deficiences contribute to Alzheimer disease pathology in rodents. Biochimica et Biophysica Acta (BBA) - Molecular Basis of Disease. 2015;**1852**(9):1687-1699

[51] Liu Z, Patil IY, Jiang T, Sancheti H, Walsh JP, Stiles BL, et al. High-fat diet induces hepatic insulin resistance and impairment of synaptic plasticity. PLoS One. 2015;**10**(5):e0128274

[52] Mody N, Agouni A, McIlroy GD, Platt B, Delibegovic M. Susceptibility to diet-induced obesity and glucose intolerance in the APP SWE/PSEN1 A246E mouse model of Alzheimer's disease is associated with increased brain levels of protein tyrosine phosphatase 1B (PTP1B) and retinol-binding protein 4 (RBP4), and basal phosphorylation of S6 ribosomal protein. Diabetologia. 2011;**54**(8):2143-2151

[53] Motor S, Ozturk S, Ozcan O, Gurpinar AB, Can Y, Yuksel R, et al. Evaluation of total antioxidant status, total oxidant status and oxidative stress index in patients with alopecia areata. International Journal of Clinical and Experimental Medicineis. 2014;7(4): 1089

[54] Nuzzo D, Picone P, Baldassano S, Caruana L, Messina E, Marino Gammazza A, et al. Insulin resistance as common molecular denominator linking obesity to Alzheimer's disease. Current Alzheimer Research. 2015;**12**(8):723-735

[55] Vargas-Robles H, Rios A, Arellano-Mendoza M, Escalante BA, Schnoor M. Antioxidative diet supplementation reverses high-fat diet-induced increases of cardiovascular risk factors in mice. Oxidative Medicine and

Cellular Longevity. 2015;**2015**:9. Article ID: 467471. DOI: 10.1155/2015/467471

[56] Timmermans S, Bogie JF, Vanmierlo T, Lütjohann D, Stinissen P, Hellings N, et al. High fat diet exacerbates neuroinflammation in an animal model of multiple sclerosis by activation of the renin angiotensin system. Journal of Neuroimmune Pharmacology. 2014;**9**(2):209-217

[57] Abbasi-Oshaghi E, Mirzaei F, Pourjafar M. NLRP3 inflammasome, oxidative stress, and apoptosis induced in the intestine and liver of rats treated with titanium dioxide nanoparticles: in vivo and in vitro study. International Journal of Nanomedicine. 2019;**14**:1919

[58] Erdogan HK, Bulur I, Kocaturk E, Saracoglu ZN, Alatas O, Bilgin M. Advanced oxidation protein products and serum total oxidant/antioxidant status levels in rosacea. Advances in Dermatology and Allergology/Postępy Dermatologii i Alergologii. 2018;**35**(3):304-308

[59] Arendt T. Synaptic plasticity and cell cycle activation in neurons are alternative effector pathways: The 'Dr. Jekyll and Mr. Hyde concept'of Alzheimer's disease or the yin and yang of neuroplasticity. Progress in Neurobiology. 2003;**71**(2-3):83-248

[60] Grillo C, Piroli G, Wood G, Reznikov L, McEwen B, Reagan L. Immunocytochemical analysis of synaptic proteins provides new insights into diabetes-mediated plasticity in the rat hippocampus. Neuroscience. 2005;**136**(2):477-486

[61] Hashimotodani Y, Nasrallah K, Jensen KR, Chávez AE, Carrera D, Castillo PE. LTP at hilar mossy cell-dentate granule cell synapses modulates dentate gyrus output by increasing excitation/inhibition balance. Neuron. 2017;**95**(4):928-943 e3

[62] Karimi SA, Komaki A, Salehi I, Sarihi A, Shahidi S. Role of group II metabotropic glutamate receptors (mGluR2/3) blockade on long-term potentiation in the dentate gyrus region of hippocampus in rats fed with high-fat diet. Neurochemical Research. 2015;**40**(4):811-817

[63] Bruel-Jungerman E, Veyrac A, Dufour F, Horwood J, Laroche S, Davis S. Inhibition of PI3K-Akt signaling blocks exercise-mediated enhancement of adult neurogenesis and synaptic plasticity in the dentate gyrus. PLoS One. 2009;**4**(11):e7901

[64] Maag JL, Panja D, Sporild I, Patil S, Kaczorowski DC, Bramham CR, et al. Dynamic expression of long noncoding RNAs and repeat elements in synaptic plasticity. Frontiers in Neuroscience. 2015;**9**:351

[65] Jo J, Whitcomb DJ, Olsen KM, Kerrigan TL, Lo S-C, Bru-Mercier G, et al. Aβ1-42 inhibition of LTP is mediated by a signaling pathway involving caspase-3, Akt1 and GSK-3β. Nature Neuroscience. 2011;**14**(5):545-547

[66] Olsen KM, Sheng M. NMDA receptors and BAX are essential for Aβ impairment of LTP. Scientific Reports. 2012;**2**(1):1-5

[67] Porter D, Faivre E, Flatt PR, Hölscher C, Gault VA. Actions of incretin metabolites on locomotor activity, cognitive function and in vivo hippocampal synaptic plasticity in high fat fed mice. Peptides. 2012;**35**(1):1-8

[68] Mihara S, Shibamoto T. The role of flavor and fragrance chemicals in TRPA1 (transient receptor potential cation channel, member A1) activity associated

with allergies. Allergy, Asthma and Clinical Immunology. 2015;**11**(1):1-12

[69] Di Pasqua R, Mamone G, Ferranti P, Ercolini D, Mauriello G. Changes in the proteome of salmonella enterica serovar Thompson as stress adaptation to sublethal concentrations of thymol. Proteomics. 2010;**10**(5):1040-1049

[70] Aman S, Moin S, Owais M, Siddiqui M. Antioxidant activity of thymol: Protective role in AAPH-induced hemolysis in diabetic erythrocytes. IJPSI. 2013;**2**(3):55-60

[71] Esmaeili A, Khodadadi A. Antioxidant activity of a solution of thymol in ethanol. Zahedan Journal of Research in Medical Sciences. 2012;**14**(7)

[72] Riella K, Marinho R, Santos J, Pereira-Filho R, Cardoso J, Albuquerque-Junior R, et al. Anti-inflammatory and cicatrizing activities of thymol, a monoterpene of the essential oil from Lippia gracilis, in rodents. Journal of Ethnopharmacology. 2012;**143**(2):656-663

[73] Jukic M, Politeo O, Maksimovic M, Milos M, Milos M. In vitro acetylcholinesterase inhibitory properties of thymol, carvacrol and their derivatives thymoquinone and thymohydroquinone. Phytotherapy Research. 2007;**21**(3):259-261

[74] Saravanan S, Pari L. Role of thymol on hyperglycemia and hyperlipidemia in high fat diet-induced type 2 diabetic C57BL/6J mice. European Journal of Pharmacology. 2015;**761**:279-287

[75] Haque MR, Ansari S, Najmi A, Ahmad MA. Monoterpene phenolic compound thymol prevents high fat diet induced obesity in murine model. Toxicology Mechanisms and Methods. 2014;**24**(2):116-123

[76] Azizi Z, Ebrahimi S, Saadatfar E, Kamalinejad M, Majlessi N. Cognitive-enhancing activity of thymol and carvacrol in two rat models of dementia. Behavioural Pharmacology. 2012;**23**(3):241-249

[77] Majlessi N, Choopani S, Kamalinejad M, Azizi Z. Thymol as a main constituent of Zataria multiflora boiss: Essential oil attenuates amyloid β-induced cognitive deficits in a rat model of Alzheimer's disease. Alzheimer's & Dementia. 2011;**7**(4):S770

[78] Youdim KA, Deans SG. Effect of thyme oil and thymol dietary supplementation on the antioxidant status and fatty acid composition of the ageing rat brain. The British Journal of Nutrition. 2000;**83**(1):87-93

Section 5

Effects of Hippocampal Lesions

Chapter 8

Subtemporal Multiple Hippocampal Transection with/without CA1-Subiculum Disconnection for Medically Intractable Temporal Lobe Epilepsy

Tomokatsu Hori, Hideki Shiramizu and Hajime Miyata

Abstract

Transsylvian selective amygdalohippocampectomy resulted in postoperative verbal memory decline in patients with mesial temporal lobe epilepsy of the language-dominant side. Mapping whole-brain connectivity changes have been studied recently of different surgical resection approaches for temporal lobe epilepsy. The subtemporal resection is the least disruptive to long-range connectivity, which may explain its better cognitive outcome. Finally, the authors introduced subtemporal multiple hippocampal transections technique in a case of hippocampal sclerosis negative left mesial temporal lobe epilepsy, and postoperative neuropsychological examinations revealed improvement of cognitive function immediately after the operation contrasting transsylvian multiple hippocampal transections in which verbal memory remains dropped. The authors introduced another new operation to left mesial temporal lobe epilepsy patient with hippocampal sclerosis by multiple hippocampal transections plus disconnection between CA1 and subiculum at the hippocampal head. Operative result is satisfactory in terms of neuropsychological and operative outcome.

Keywords: amygdalohippocampectomy, transsylvian, subtemporal, multiple hippocampal transection (MHT), disconnection CA1 and subiculum, neuropsychological outcome, hippocampal sclerosis (HS)

1. Introduction

The limbic lobe is located at the most medial portion of the cerebral hemisphere, like as a band surrounding the orifice into the lateral ventricle. The limbic lobe is mainly consisted of the hippocampal formation, the amygdaloid complex, and the cingulate cortices. These areas are concerned with the basic brain higher functions, such as emotion, memory, attention, cognition, and so on. Each area has the specific structural organization and the fiber connections executing the specific function. The circuit of memory, so-called Papez circuit, is consisted of the

IntechOpen

hippocampal formation, mammillary body, anterior and midline thalamic nuclei, posterior cingulate cortex, and the retrohippocampal cortices. On the other hand, the circuit of the emotion, so-called Yakovlev circuit, is consisted of the amygdala, the mediodorsal thalamic nucleus, anterior cingulate cortex, and the orbitofrontal cortex. Although memory and emotion are processed on the independent circuit, there are also several structures where the fibers from the two circuits meet together, such as the nucleus accumbens, the entorhinal cortex, and the hypothalamic area. Memory information is stored efficiently only when the brain is in appropriate state for memory acquisition. This state, the motivating state for memory, is determined by the levels of awareness, cognition, attention, emotion, and other influences. The motivating state might change signal processing in the hippocampus. Classical anatomical observations with Golgi staining of hippocampal neurons are the basis for the simple trisynaptic circuit (dentate granule cells (DG)-Cornu Ammonis 3 (CA3), CA3 to CA1 concept of hippocampal function (**Figure 1**).

Recent anatomical work has revealed much richer synaptic connections between hippocampal neuron subfields (DG, CA3, CA2, and CA1) and wide distribution of axons along the longitudinal axis of the hippocampus [1].

Temporal lobe epilepsy (TLE) is involving the limbic system, especially amygdala and hippocampus which influences emotion, memory, attention, and cognition.

Despite various medical drugs have been tried to control intractable TLE, surgical treatment gives better seizure control comparing to prolonged medical

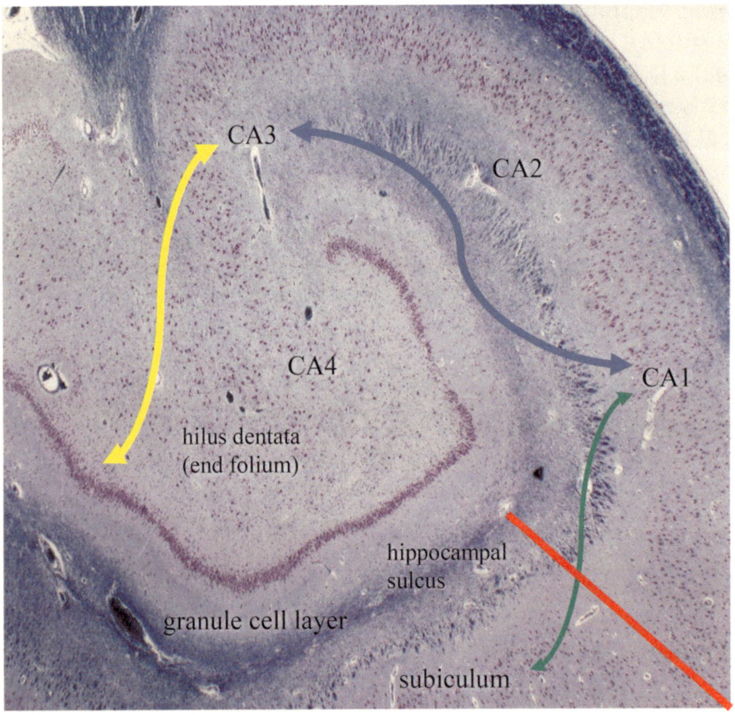

Figure 1.
Schema illustrating disconnection between CA1 and subiculum (red line). Yellow arrow is mossy fiber, blue arrow is Schaffer collateral, green arrow is alveus hippocampi, and red line is disconnection between CA1 and subiculum just until hippocampal sulcus. This histology is sectioned from normal hippocampus taken from cadaver without central nervous lesion.

treatment [2]. But in this randomized control study, substantial number (around 5%) of surgical patients complained of postoperative memory difficulty, but the authors acknowledge that although this complication is important, but the benefit is worth the risk. Concerning such memory problems, in surgical population, MR negative language-dominant-side mesial TLE group, so-called paradoxical temporal lobe epilepsy (PTLE) [3], is at great risk of postoperative memory deficits. Until now, neurosurgeons have made great efforts to stop such a postoperative memory decline by adopting various operative approaches, such as reducing size of temporal lobectomy, selective amygdalohippocampectomy, and recent radiosurgical interventions [4].

In this chapter of this book, our further efforts to escape from such a postoperative memory decline such as subtemporal multiple hippocampal transection (MHT) with or without disconnection between cornu ammonis (CA1) and subiculum will be presented in detail (**Figure 1**). These techniques and results are never reported in the world literature.

2. Neuropsychological results of various surgical treatments for TLE

Temporal lobe epilepsy (TLE) is one of the most intractable epilepsy involving the limbic system, especially amygdala and hippocampus which influences emotion, memory, attention, and cognition.

Despite various medical drugs have been tried to control intractable TLE, surgical treatment gives better seizure control comparing to prolonged medical treatment [2]. Concerning memory problems, in surgical population, MR negative language-dominant-side mesial TLE group, so-called paradoxical temporal lobe epilepsy (PTLE) [3], is at great risk of postoperative memory deficits. Until now, neurosurgeons have made great efforts to stop such a postoperative memory decline by adopting various operative approaches, such as reducing size of temporal lobectomy, selective amygdalohippocampectomy, and recent radiosurgical interventions [4].

Morino et al. reported comparison of neuropsychological results after selective amygdalohippocampectomy versus anterior temporal lobectomy (ATL) [5]. Transsylvian selective amygdalohippocampectomy (TSSAH) is an operative technique planned to spare unaffected brain region during surgical treatment for mesial temporal lobe epilepsy (MTLE). In contrast to standard anterior temporal lobectomy (ATL), the advantages of TSSAH with respect to postoperative cognitive outcome are equivocal without randomized control study. Morino et al. compared cognitive function before and after surgery in 49 patients with unilateral mesial temporal lobe seizures who underwent either ATL (n = 17) or TSSAH (n = 32). All patients received neuropsychological testing before and 1 year after surgery. The intelligence quotient (IQ) increased postoperatively in both surgical groups. Memory evaluation in the ATL group revealed a postoperative decline in nonverbal memory after right-sided resection and a postoperative decline in verbal memory after left-sided resection. In the TSSAH group, there was a slight postoperative decline only in verbal memory after left-sided resection, but other memory function was preserved. There was significant postoperative improvement in verbal memory after right-sided resection. Overall, memory function was better preserved in the TSSAH group than in the ATL group.

There is another change of operative procedure invented by Shimizu et al. [6], that is, transsylvian multiple hippocampal transection (TSMHT).

Table 1 is a summary of memory scores by transsylvian multiple hippocampal transection (TSMHT) for PTLE [6, 7]. Preoperative verbal memory score dropped 1 month after surgery, but 1 year after surgery improved up to preoperative level.

How about the neuropsychological results after subtemporal **amygdalohippocampectomy (sSAH)** for **TLE** [8, 9]. **Figure 2** shows combined neuropsychological results of sSAH by Takaya (left panel) [8] and Hori (right panel) [9]. Both WMS-R and WAIS-R scores show significant improvements postoperatively comparing to preoperative levels. This is the difference between TSSAH and sSAH.

Usami et al. reported their operative results by transsylvian approach with multiple hippocampal transection and multiple subpial transection (MST) with lesionectomy (TSMHT +MST/L) technique [10] for TLE. As it is clearly shown, follow-up results show some deterioration of scores, especially VIQ. **Figures 2** and **3** show the difference of postoperative cognitive function between TSSAH MHT + MST/L and sSAH [8–10].

There were **chang**es in neuropsychological function after surgery on the verbally dominant side (n = 12). There were no significant differences between the preoperative indices and those at the last v**isi**t. The values were as follows (mean $\pm$ SD preoperatively, at the last v**isi**t; p-value): verbal memory (85 $\pm$ 13, 78 $\pm$ 18; 0.14), visual memory (94 $\pm$ 24, 102 $\pm$ 16; 0.08), general memory (85 $\pm$ 16, 82 $\pm$ 19; 0.29), delayed recall (79 $\pm$ 18, 87 $\pm$ 19; 0.09), attention and concentration (89 $\pm$ 21, 88 $\pm$ 22; 0.10). Regarding IQ, there was a significant difference in VIQ (87 $\pm$ 19, 80 $\pm$ 19; 0.045*), but not in PIQ (89 $\pm$ 28, 88 $\pm$ 23; 0.42) or FIQ (86 $\pm$ 25, 82 $\pm$ 21; 0.16) [10].

3. Subtemporal amygdalohippocampectomy

There are many surgical techniques to cure medically intractable TLE such as conventional temporal lobectomy. But, concerning the language-dominant-side TLE without hippocampal sclerosis (PTLE) surgical removal of mesial temporal structures may result in neuropsychological problems, especially decline of verbal memory. Usually, temporal lobectomy 4–4.5 cm away from the temporal tip is used for language-dominant-side TLE [2]. Postoperative verbal memory decline is a major concern especially for language-dominant-side TLE without HS so-called PTLE [3, 7]. For mesial TLE of which amygdala and hippocampal head are epileptic foci, selective amygdalohippocampectomy is used by various routes including transsylvian, through superior temporal gyrus or sulcus (T1), middle temporal gyrus or sulcus (T2), and inferior temporal gyrus or sulcus (T3) depending on the preference of neurosurgeons (**Figure 4**). In 1993, subtemporal amygdalohippocampectomy (sSAH) technique has been introduced for mesial temporal lobe epilepsy to abolish postoperative neuropsychological deterioration observed in usual anterior temporal lobectomy [11]. The conventional subtemporal approach has been modified to diminish temporal lobe compression pressure and the risk of damage to the temporal stem. In this technique, the approach has been changed from usual anterolateral approach to posterolateral, thereby avoiding the voluminous and deeply embedded anterior temporal lobe in the middle fossa. By this approach, the retraction pressure is decreased and the temporal stem which is important bottleneck of temporal information were spared. To

	Lt (n = 15)						Rt (n = 11)					
	Preop.	**1MP**	**1YP**	**Preop.-1MP**	**Preop.-1YP**	**1MP-1YP**	**Preop.**	**1MP**	**1YP**	**Preop.-1MP**	**Preop.-1YP**	**1MP-1YP**
Verbal	75.7 (4.8)	73.2 (4.9)	75.0 (6.3)	p = 0.3257	p = 0.8754	p = 0.7069	77.3 (5.5)	81.3 (6.0)	83.8 (6.1)	p = 0.1887	p = 0.0970	p = 0.5225
Performance	77.7 (6.3)	81.4 (6.3)	79.6 (7.5)	p = 0.2547	p = 0.7187	p = 0.7293	77.1 (7.1)	84.4 (7.2)	91.8 (7.8)	p = 0.0794	†p = 0.0071	p = 0.1340
Total	74.0 (5.8)	74.3 (5.9)	73.9 (6.8)	p = 0.8922	p = 0.9875	p = 0.9214	74.4 (6.6)	79.4 (6.7)	87.3 (7.2)	p = 0.1262	†p = 0.0033	p = 0.0603
No significant improvement or decline are observed for verbal intelligent quotient (IQ), performance IQ, and total IQ.												
Verbal memory (WMS-R) Morino et al. [7]												
	Preop.	1MP	1YP	Preop.-1MP	Preop.-1YP	1MP-1YP	Preop.	1MP	1YP	Preop.-1MP	Preop.-1YP	1MP-1YP
	86.1 (5.9)	71.3 (6.2)	79.8 (6.0)	*p = 0.04	p = 0.1845	p = 0.0966	89.1 (6.2)	90.8 (6.2)	97.3 (6.3)	p = 0.7165	p = 0.1009	p = 0.1922

1MP: 1 month after op. 1YP: 1 year after op.
Only WMS-R verbal memory is deteriorated 1 month after operation, but it recovered to preoperative level 1 year after the operation, but never statistically improved comparing to preoperative level.

Table 1.
Memory outcomes depending on the operative side by transsylvian hippocampal transection for PTLE by Morino et al. [7].

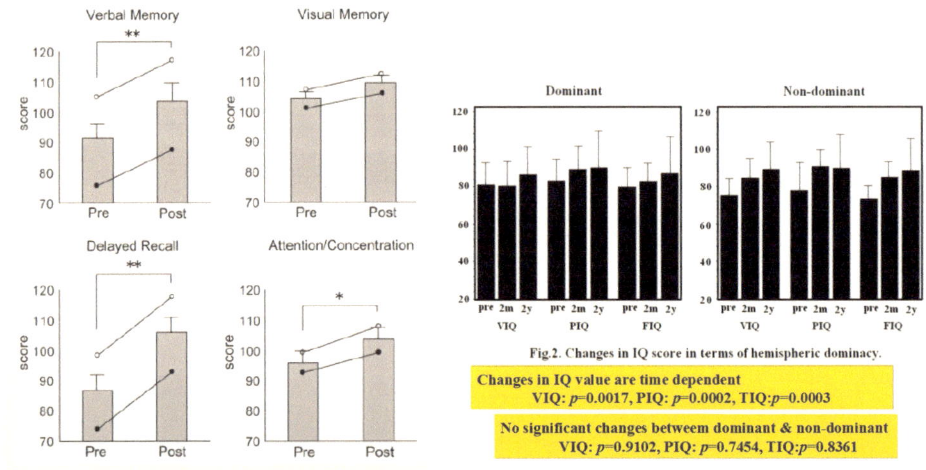

Figure 2.
Left panel adopted from Takaya et al. [8], and right panel from Hori et al. [9]. Left panel: cognitive improvement might result from a combined effect of good seizure control and minimize the regions of the brain with postoperative functional impairment. Improved cerebral function in terms of WMS-R scores in mesial temporal lobe epilepsy after sSAH was demonstrated [8]. Right panel: bar graphs showing changes in IQs in patients in whom the language-dominant and nondominant hemisphere was resected by sSAH [9].

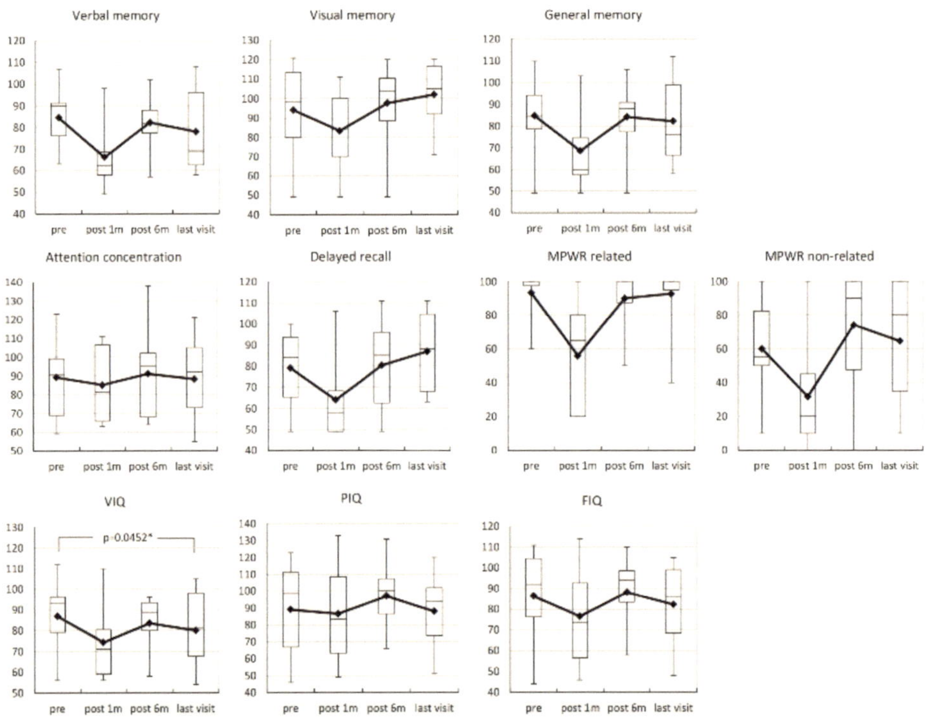

Figure 3.
Adopted from Usami et al. [10].

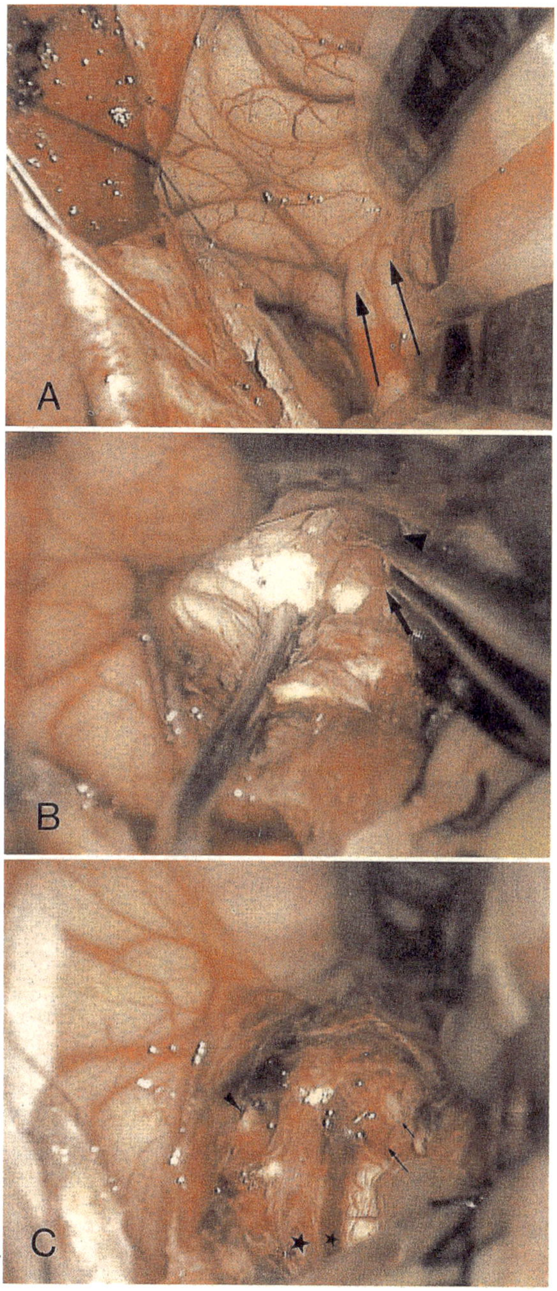

Figure 4.
A,B,C Microscopic procedure of subtemporal amygdalohippocampectomy. A: Temporal base is retracted, and fusiform gyrus is identified (double arrows). A small bridging vein was just riding on the collateral sulcus, and the vein is preserved. B: Fusiform gyrus, collateral sulcus is identified, and temporal horn is opened, then amygdala and hippocampal head are identified. At the choroid plexus (arrow head), the anterior choroid artery (arrow) is identified at the tips of the forceps. The suction tube touches the surface of the hippocampus. After the hippocampus (about 2.5 ~ 3.0 cm usually in length) anterior to the choroidal point is removed, the amygdala, the parahippocampal gyrus, and the uncus were removed, so that in every patient the amygdala can be removed en bloc, *and neuropathologically is examined. It is different point between TSSAH and sSAH. C: After sSAH, PCA (black star), basal vein (small black star), optic nerve (short arrow), and anterior choroidal artery (long arrow) are well exposed.*

date, the authors using this approach have operated on more than 50 patients with medically intractable temporal lobe epilepsy whose epileptic foci were in the mesial temporal lobe structure; the inferior temporal gyrus, the temporal tip, the vein of Labbe, and the ventral bridging veins were preserved using with dissecting technique without adverse events. It can be used to remove as much of the posterior hippocampus as necessary, and it can be extended to conventional lobectomy if necessary. If there is some drawback in this approach, a part of basal temporal language area is sacrificed in order to reach temporal horn (**Figure 5A-C**). One patient among 50 patients with HS negative language-dominant-side TLE (PTLE) has shown postoperative severe memory deficits. Although the patient has shown gradual improvement of her memory in these 20 years during her postoperative period, she is still complaining of memory problem and it is compromising her daily job, even if operated upon by sSAH.

4. Minimally invasive subtemporal approach

Recently, Busby et al. [12] reported whole-brain tract changes after mapping and analyzed the potential impact of different surgical resection approaches for TLE. The main aim of this study was to perform systematic "pseudo-neurosurgery" based on existing resection methods on healthy neuroimaging data and measuring the effect on long-tract connectivity. They use anatomical connectivity mapping (ACM) to determine long-range disconnection, which is complementary to existing measures of local integrity such as fractional anisotropy or mean diffusivity. ACMs were generated for each diffusion scan in order to compare whole-brain connectivity with an "ideal resection," nine anterior temporal lobectomy and three selective approaches. For *en bloc* resections, as distance from the temporal pole increased, reduction in connectivity was evident within the arcuate fasciculus, inferior longitudinal fasciculus, inferior front-occipital fasciculus, and the uncinate fasciculus.

Increasing the height of resections dorsally reduced connectivity within the uncinate fasciculus. sSAH was associated with connectivity modes most similar to the "ideal" baseline resection, compared to TSSAH and middle-temporal approaches. In conclusion, Busby N, et al. showed the utility of ACM in assessing long-range disconnections/disruptions during temporal lobe resections, where they identified the sSAH as the least disruptive to long-range connectivity which may explain its better cognitive outcome. Of course, magnetic resonance (MR)-guided focused ultrasound treatment of mesial TLE is an ideal treatment if properly sonicated at the key structures of intractable TLE.

In 2021, Whiting AC and Smith KA, et al. [13] reported seizure and neuropsychological outcomes in a large series of selective amygdalohippocampectomies with a minimally invasive sSAH which is almost similar to our approach.

All patients in this study had at least 1 year of follow-up (mean [SD] 4.52 [2.57] years), of whom 57.9% (88/152) had Engel Class I seizure outcomes.

Engel's classification is as follows:

Class I: Seizure free or no more than a few early, nondisabling seizures; or seizures upon drug withdrawal only.

Class II: Disabling seizures occur rarely during a period of at least 2 years; disabling seizures may have been more frequent soon after surgery; nocturnal seizure.

Class III: Worthwhile improvement; seizure reduction for prolonged periods but less than 2 years.

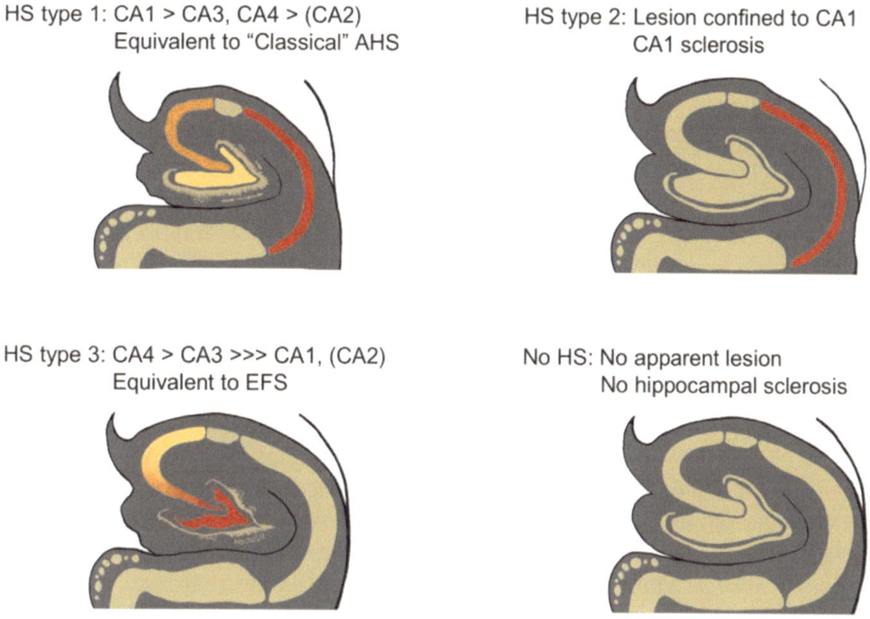

Figure 5.
Types and Frequency of HS in our operative specimen in terms of ILAE Classification. HS type 1 (25 of 41 cases, 61%) is equivalent to 'classical' Ammon's horn sclerosis in which neuronal loss and gliosis is most severe in CA1, followed by CA3, CA4, with relative sparing of CA2 and often associated with loss of dentate granule cells and/or dispersion. HS type 2 represents neuronal loss and gliosis almost confined to CA1 (CA1 sclerosis), and only 1 case (2%) was identified in our study. HS type 3 (7 cases, 17%) is characterized by a reverse distribution of the sclerotic lesion to HS type 1, in which neuronal loss and gliosis is the most severe in CA4 followed by CA3, with relative sparing of CA2 and CA1, that is equivalent to endofolium sclerosis (EFS). In addition to these three HS types, we also identified 8 cases (19%) without apparent neuronal loss and gliosis (no HS).

Class IV: No worthwhile improvement; some reduction, no reduction, or worsening are possible.

The patients with at least 2 years of follow-up (mean [SD] 5.2 [2.36] years), 56.5% (70/124) had Engel Class I seizure outcomes. Of the 152 patients with at least 1 year of clinical follow-up, only 38 (25%) completed both preoperative and postoperative neuropsychological testing by a neuropsychologist. The mean difference in scores was statistically significant in the dominant hemisphere group for the RAVLT–short delay test ($p = 0.02$) and the BNT (Boston Naming Test) ($p = 0.04$). The mean difference in scores was statistically significant in the nondominant group only with the BNT ($p = 0.04$). Many patients were unable to complete both preoperative and postoperative neuropsychological examinations for a variety of reasons. Concerning this decline, it might be influenced by some bias, that is, only 25% of all patients are examined, and a small number of neuropsychologically declined group of patients are prone to be examined.

Adverse events were low, with a 1.3% (2/152) permanent morbidity rate and 0.0% mortality rate.

This study reports a large series of patients who have undergone sSAH, with a minimally invasive technique. The sSAH approach described in this study appears to be a safe, effective, minimally invasive technique for the treatment of MTLE among the surgical methods ever reported in the literature.

5. Neuropsychological results after subtemporal amygdalohippocampectomy

In 2007, the authors [9] evaluated operative, neuropathological, and neuropsychological results after selective subtemporal amygdalohippocampectomy (sSAH) for refractory temporal lobe epilepsy in patients who were observed for at least 2 years after surgery. More than 26 consecutive patients underwent sSAH for non-lesional, medically refractory TLE. Neuropsychological evaluation using the Wechsler Adult Intelligence Scale (WAIS) was done before surgery in all patients, 2 months after surgery in 24 patients, and at 2-year follow-up in 19 patients. The data were compared between the 13 patients in whom the language-dominant hemisphere was surgically treated and the six patients in whom the language-nondominant hemisphere was treated. After surgery, 84% of the patients showed either Engel Class I or II seizure outcome. There were no permanent subjective complications except postoperative memory impairment in one patient with normal intelligence without HS. Neuropathological examination revealed HS in 19 patients. No significant differences in IQ and verbal memory test scores were observed between the patients in whom the language-dominant hemisphere was treated and those in whom the language-nondominant hemisphere was treated. One patient without HS whose language-dominant hemisphere was treated by sSAH, postoperative memory loss is only her complication. In this patient, although her memory has slightly improved, her job is compromised by memory loss. Considering this situation, neurosurgeons should explore better operative technique to escape from such postoperative memory loss, especially for language-dominant-side TLE without HS (PTLE).

Significant postoperative improvements in verbal IQ, performance IQ (PIQ), and full-scale IQ (FIQ) were observed over time. No significant differences were found between pre- and postoperative verbal memory test scores, and no subjective visual field loss was marked in any patient. Thus, sSAH provides good surgical and neuropsychological results and does not cause significant postoperative decline of verbal memory even if performed on the language-dominant side. In **Figure 2**, graphs depicting changes in IQs in patients in whom the language-dominant hemisphere was resected and patients in whom the language-nondominant hemisphere was resected, changes were time-dependent (VIQ, $P = 0.0107$; PIQ, $p = 0.0002$; FIQ, $p = 0.0003$), with no significant differences between the dominant and nondominant hemisphere groups (VIQ, $p = 0.9102$, PIQ, $p = 0.7454$; FIQ, $p = 0.8361$), and significant increases in VIQ, PIQ, and FIQ were observed over time.

Takaya et al. [8] evaluated the effects of sSAH on cerebral glucose metabolism and memory function in 15 patients with medically intractable MTLE with HS using [18F]-fluorodeoxyglucose PET (FDG-PET) and the Wechsler Memory Scale-Revised (WMS-R). The patients were evaluated before and 1–5 years (mean 2.6 years) after surgery. In patients with MTLE of the language-dominant hemisphere, the basal temporal language area was preserved by this surgical approach. Postoperative glucose metabolism increased in extratemporal areas ipsilateral to the affected side, such as the dorsolateral prefrontal cortex, and the dorsomedial and ventromedial frontal cortices. Glucose metabolism also increased in the bilateral inferior parietal lobules and in the remaining temporal lobe regions remote from the resected mesial temporal region, such as the superior temporal gyrus and the temporal pole. By contrast, postoperative glucose metabolism decreased only in the mesial temporal area near the resected region. Postoperative verbal memory, delayed recall, and attention/concentration scores were significantly better than preoperative scores regardless of the resected side (**Figure 2** Left). This study suggests that the selective removal of the epileptogenic region in MTLE using subtemporal approach improved cerebral glucose metabolism in the areas receiving projections from the affected mesial temporal lobe. Cognitive improvement might result from a combination of good seizure control and minimize the area of the brain with postoperative functional impairment. Improved cerebral function in terms of WMS-R scores in mesial temporal lobe epilepsy after sSAH was demonstrated in **Figure 2** Left [8].

Judging from these figures, the difference between our neuropsychological results and Takaya's results is not clear, indicating that sparing incision of basal language area in Takaya's series may not influence the results.

6. Histological classification of hippocampal sclerosis

As demonstrated in **Figure 5**, HS type 1 (25 of 41 cases, 61%) is equivalent to "classical" Ammon's horn sclerosis in which neuronal loss and gliosis are most severe in CA1, followed by CA3, and CA4, with relative sparing of CA2 and often associated with loss of dentate granule cells and/or dispersion [14, 15]. HS type 2 represents neuronal loss and gliosis almost confined to CA1 (CA1 sclerosis), and only one case (2%) was identified in our study. HS type 3 (7 cases, 17%) is characterized by a reverse distribution of the sclerotic lesion to HS type 1, in which neuronal loss and gliosis are the most severe in CA4 followed by CA3, with relative sparing of CA2 and CA1, that is equivalent to endofolium sclerosis (EFS). In addition to these three HS types, we also identified eight cases (19%) without apparent neuronal loss and gliosis (no HS). Subiculum was relatively well preserved in all cases. Granule cell dispersion is one of the abnormal structural changes that has been shown in patients with temporal lobe epilepsy. In a normal situation, the granule cells in dentate gyrus should be tightly packed. But in granule cell dispersion, the compact formation was lost, and the axons need to extend longer to reach the neighboring granule cells. It might be a consequence of a migration disorder, and the first hypothesis considers an initial injury that releases toxin(s) that affect the normal migration of granule cells. The second hypothesis concerns the role of reelin. Reelin is required for normal neuronal lamination in humans, and the lack of this expression can lead to migration defect associated with temporal lobe epilepsy.

Types of HS did not correlate with age at operation and duration of illness, suggesting that these types represent distinct pathology of MTLE, the mean age of onset in patients with type 1 sclerosis tends to be younger than those at least with no HS but this is not statistically significant (Kruskal-Wallis test), the history of initial precipitating injury is not correlated with histological subtypes or postoperative seizure control, and type 1 sclerosis seems to correlate with better postsurgical seizure outcome than other types [14, 15].

The choice of the operative procedure is important factor affecting the seizure outcome, and that lateral temporal structure is also involved in the epileptogenicity in a subset of patients with MTLE (**Tables 2** and **3**).

In 2019, Seki et al. [16] reported an analysis of proliferating neuronal progenitors and immature neurons in the human hippocampus surgically removed from control and epileptic patients. Adult neurogenesis in the mammalian hippocampus is a well-known phenomenon (**Figure 6**).

HS types	N (%)	Ammon's horn on MRI (%)			Ammon's horn pathology (%)			Amygdala pathology (%)					Dual pathology (%)
		n	HI-signal	Atrophy	n	N-swell	GCD	n	N-swell	DG	FG	NL & G	
Type 1	25 (61.0)	25	23 (92.0)	19 (76.0)	25	24 (96.0)	24 (96.0)	21	13 (61.9)	17 (81.0)	1 (4.8)	1(4.8)	6125 (24.0)
Type 2	I (2.4)		1	0		0	1, focal		0		0	0	II 1
Type 3	7 (17.1)	7	3 (42.9)	3 (42.9)	7	1 (14.3)	4 (57.1)	6	5 (83.3)	6 (100)	0 (0.0)	0 (0.0)	3/7 (42.9)
No HS	8(19.5)	7	4 (57.1)	3 (42.9)	8	2 (25.0)	1 (12.5)	8	5 (62.5)	7 (87.5)	1 (12.5)	0 (0.0)	—
Total	41 (100)	40	31 (77.5)	25 (62.5)	41	27 (65.9)	30 (73.2)	36	23 (63.9)	31(86.1)	2 (5.6)	I (2.8)	10,133 (30.3)

Abbreviations: DG: diffuse gliosis without significant neuronal loss, FG: focal gliosis, GCD: granule cell dispersion; HI-signal: high-intensity signal on T2-weighted and/or FLAIR image, N-swell: swelling of remaining neurons; NL & G: neuronal loss and gliosis.

Table 2.
*ILAE HS type 1 is found in 25/41 (61%) patients, while amygdala sclerosis is found in only one patient (**Figure 5**), and dual pathology is found in 10/33 (30.3%).*

HS types	N (%)	Onset (y) Mean ± SD	Duration (y) Mean ± SD	IPI (%)	Engel's class (n)				
					I	II	III	IV	n/a
Type 1	25 (61.0)	12.0 ± 11.0	19.7 ± 10.1	17/25 (68.0)	19	2	4	0	
Type 2	I (2.4)	29	21	0/1			1		
Type 3	7 (17.1)	17.9 ± 14.0	16.1 ± 9.7	2/7 (28.6)	3	1	2	1	
No HS	8 (19.5)	22.1 ± 10.0	15.4 ± 13.2	4/6 (66.7)	5	1	1	0	1
Total	41 (100)	14.7 ± 11.7	18.4 ± 10.4	22/37 (59.5)	27	4	8	1	1

Abbreviations: IPI: clinical history of initial precipitating injury; SD: standard deviation.

Table 3.
HS types and clinical features.

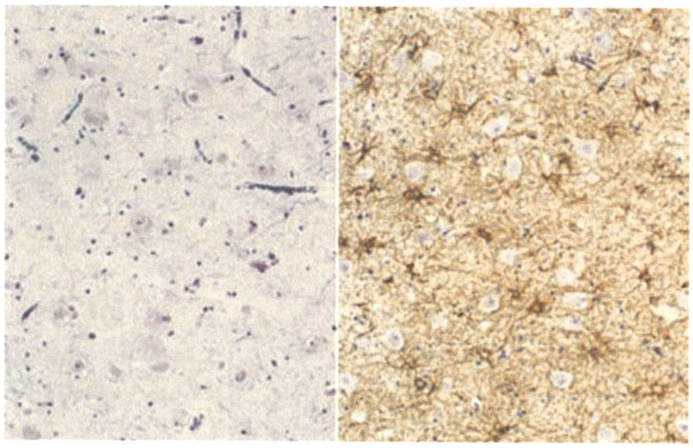

Figure 6.
*Amygdala sclerosis. Amygdala sclerosis is detected in only one patient with HS type 1 (**Table 2**). Amygdala basolateral nucleus, Left: Nissl' stain, Right: glial fibrillary acidic protein (GFAP) stain.*

It remains controversial as to what extent adult neurogenesis actually occurs in hippocampus, and how brain diseases, such as epilepsy, affect human adult neurogenesis. We analyzed polysialylated neural cell adhesion molecule (PSA-NCAM) cells and proliferating neuronal progenitor (Ki67+/mammalian Hu protein B (HuB)+/ doublecortin (DCX) + cells in the surgically removed hippocampus of epileptic patients. In control patients, a substantial number of PSA-NCAM+ cells were distributed densely below the granule cell layer (GCL). In epileptic patients with granule cell dispersion, the number of PSA-NCAM+ cells were reduced, and aberrant PSA-NCAM+ cells were found. However, the numbers of Ki67+/HuB+/DCX+ cells were very low in both control and epileptic patients. The large number of PSA-NCAM+ cells and few DCX+/HuB+/Ki-67+ cells observed in the controls suggest that immature-type neurons are not recently generated neurons, and that the level of hippocampal neuronal production in adult humans is low. These results also suggest that PSA-NCAM is a useful marker for analyzing the pathology of epilepsy, but it is not evident that these bizarre PSA-NCAM neurons are the results or cause of intractable epilepsy (**Figure 7**). Different interpretations of the immunohistochemical results between humans and rodents should be examined in future.

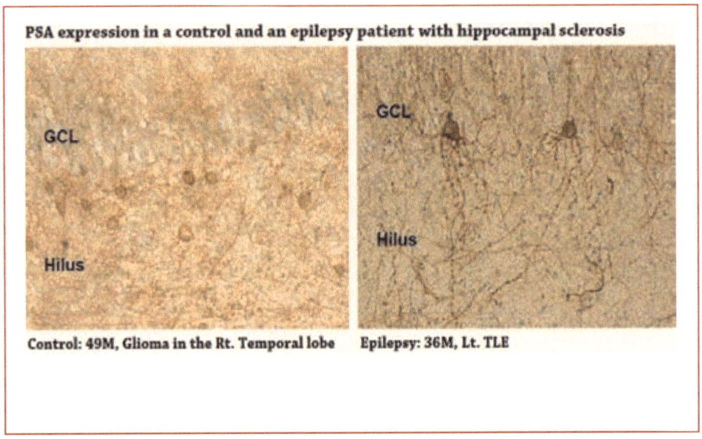

Figure 7.
In control patients, a substantial number of PSA-NCAM+ cells were distributed densely below the granule cell layer (GCL). In epileptic patients with granule cell dispersion, the number of PSA-NCAM+ cells were reduced, and aberrant PSA-NCAM+ cells were found.

7. Technique of the multiple hippocampal transection (MHT)

The consequences of resection of the hippocampus, where its function is still preserved, can be a decrease in verbal memory or visual–spatial memory, intelligence, emotional and speech performance, as well as cognitive disorders. To solve this problem, Shimizu et al. [6] in 2006 reported the technique of the multiple hippocampal transection (MHT). The concept of MHT originated on the basis of multiple subpial transection in eloquent areas of the neocortex. Uda et al. [17] reported differences based on the surgical side: MHT on the nondominant side resulted in significant improvements in verbal but not visual memory, whereas MHT on the dominant side did not lead to significant increase in verbal or visual memory.

The principle of surgical treatment of mesial temporal lobe epilepsy by multiple transverse transection of the hippocampus is the mechanical disruption of the longitudinal pathways of the hippocampus.

Recent anatomical evidence suggests a functionally significant back-projection pathway from the subiculum to the CA1. A critical role for CA1-projecting subicular neurons in object-location learning and memory show that this projection modulates place-specific activity of CA1 neurons and their responses to displaced objects. Together, these experiments reveal a novel pathway by which cortical inputs, particularly those from the visual cortex, reach the hippocampal output region CA1. It is established that the hippocampus has two types of pathways: (1) trisynaptic pathways, which are located in parallel loops oriented orthogonally to the longitudinal axis of the hippocampus; (1) from entorhinal cortex to granule cell layer (perforant fiber), (2) from GCL to CA3 (mossy fiber), and (3) from CA3 to CA1 (Schaffer collateral fiber). These fiber systems are so-called trisynaptic circuit of the hippocampus (**Figure 1**).

There are two longitudinal pathways that run along the long axis of the hippocampus [7]. Loops of trisynaptic pathways going into the entorhinal cortex are important for processing and stabilizing memory. The longitudinal path of the hippocampus does not play an important physiological role; on the contrary, it facilitates the

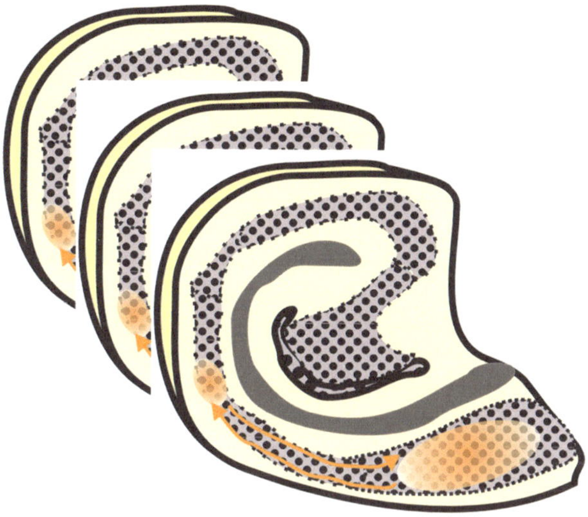

Figure 8.
Multiple subtemporal hippocampal transection every 5 mm apart.

synchronization of pathological epileptic discharges and their propagation along the hippocampus and, further, to extrahippocampal structures, thus contributing to the development of a seizure. For the pathological electrical activity in the hippocampal neurons to develop into an epileptic seizure, synchronization of the critical number of neurons – exceeding 5 mm thickness – located in the hippocampal segment is necessary. Therefore, if longitudinal horizontal interneuronal fibers along the axis of the hippocampus are separated with an interval of 5 mm, then the pathological connection is interrupted and, an epileptic seizure stop (**Figure 8**).

8. The results of long-term changes in cognitive function after MHT on the verbally dominant side

Usami et al. [10] reported the results of long-term changes in cognitive function after surgery on the verbally dominant side (n = 12). This clinical research is most recent and reliable because of containing fair number of verbally dominant side. There were no significant differences (**Figure 3**) between the preoperative indices and those at the last visit. Regarding intelligence, there was a significant difference in VIQ (8719,8019; 0.045*), but not in PIQ or FIQ. They concluded that, in all neuropsychological batteries, the average indices declined temporarily at 1 month, recovered to the preoperative level at 6 months, and were maintained for a long time after MHT + multiple subpial transection/lesionectomy (MST/L). There were no statistically significant differences between the preoperative and last-visit values in all batteries. In three patients, verbal memory indices dropped >20 points from preoperative figures after >5 years. Although VIQ (verbal intelligence quotient) and FIQ (full-scale intelligence quotient) declined temporarily at 1 month, they recovered to preoperative levels at 6 months. PIQ (performance intelligence quotient) and FIQ were preserved at the last visit, whereas VIQ had declined at the last visit in comparison with that on the preoperative test (p = 0.045). Judging from these results, transsylvian approach influenced this decline of VIQ comparing to our gain of VIQ for language-dominant side by sSAH operation (**Figure 2** Right Panel).

The important finding was that there was a significant discrepancy between memory indices and morphologic changes of the mesial temporal lobe and associated structures.

Memory preservation: postoperative cognitive impairment has been an important and controversial issue in the surgical treatment of mTLE. Patients with PTLE have a significant risk of postoperative memory decline. Long-term observation after medial temporal resection revealed that the memory impaired by surgery did not recover over time.

Although the Wada test (sodium amytal is injected into the internal carotid artery to induce a temporary state of hemianesthesia during which language and memory function of the unaffected hemisphere are tested) is not a reliable predictor of postoperative memory decline, Usami et al. recently demonstrated that parahippocampal high-gamma activity could provide predictive information about whether the mesial temporal lobe can be resected without causing memory decline.

The postoperative decline in verbal memory impairs cognitive performance in patients with MTLE. Verbal memory function after anterior temporal lobectomy or transsylvian SAH deteriorates at the group level in patients with dominant-side MTLE, whereas it tends to improve in patients with nondominant-side MTLE (Morino et al. [5]).

In our study, an improvement in verbal memory was observed regardless of the resected side. Previous studies have reported that sSAH might escape from verbal memory decline in patients with dominant-side MTLE [7, 18, 19]. Preservation of the basal temporal language area resulted in improved verbal memory 1 year after the operation, even when the anti-epileptic drug (AED) dosage remained unchanged. Mikuni et al. also shows a long-lasting improvement in verbal memory following sSAH. The basal temporal language area is located between 10 mm and 75 mm posterior to the temporal tip and is important in processing verbal information. Verbal IQ is the ability to understand and reason using concepts framed in words, and it improved after 2 years. Performance IQ (PIQ) is designed to provide a measure of an individual's overall visuospatial intellectual abilities score and full-scale IQ (FIQ) which is an overall score as well as scores for component abilities. Both PIQ and FIQ improved after both 2 months and 2 years postoperatively.

9. Subtemporal multiple hippocampal transection (New technique to preserve memory)

In 2021, a 51-year-old man showing left amygdala enlargement with medically intractable epilepsy patient without hippocampal sclerosis (HS) was introduced to our clinic. FDG-PET analysis (**Figure 9**) showed left mesial temporal lobe epilepsy with normal intelligence; in this patient, the authors adopted subtemporal selective amygdalotomy with multiple hippocampal transections (**Figures 8** and **10**). He has shown improved neuropsychological examination scores 3 months after surgery, and intractable seizure stopped after surgery (Engel's Class I) and returned to his previous job immediately after the operation (**Figure 11**, **Table 4**).

10. Pathophysiological characteristics associated with epileptogenesis in human HS

In 2017, Kitaura et al. [20] reported pathophysiological characteristics associated with epileptogenesis in human HS. Majority of seizures originate primarily from the

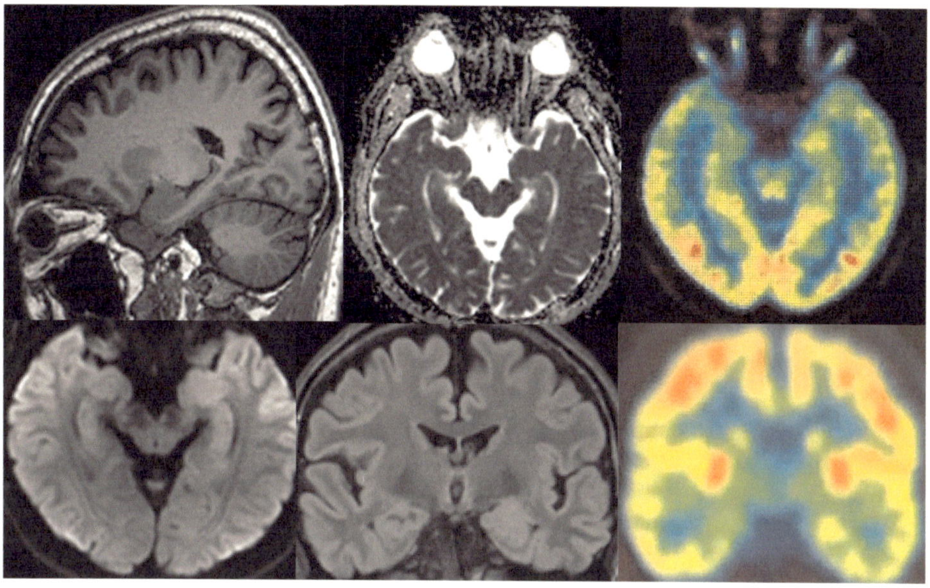

Figure 9.
MR showing left amygdala enlargement, hyperintensity (FLAIR lower left and lower middle), and FDG-PET demonstrated hyper FDG activity at the amygdala (upper & lower right).

hippocampus. They investigated epileptiform activities ex vivo using living hippocampal tissue taken from patients with MTLE. Flavoprotein fluorescence imaging and local field potential recordings revealed that epileptiform activities developed from the subiculum. Moreover, physiological and morphological experiments revealed possible impairment of K+ clearance in the subiculum affected by HS. K+ clearance is mainly regulated by astrocytes Kir 4.1 so that these findings indicate the role of astrocytes in epileptogenesis in HS. Stimulation of mossy fibers induced recurrent trans-synaptic activity in the granule cell layer of the dentate gyrus, suggesting that mossy fiber sprouting in HS also contributes to the epileptogenic mechanism presumably in addition to bizarre PSA-NCAM positive immature neurons observed in our specimen. These

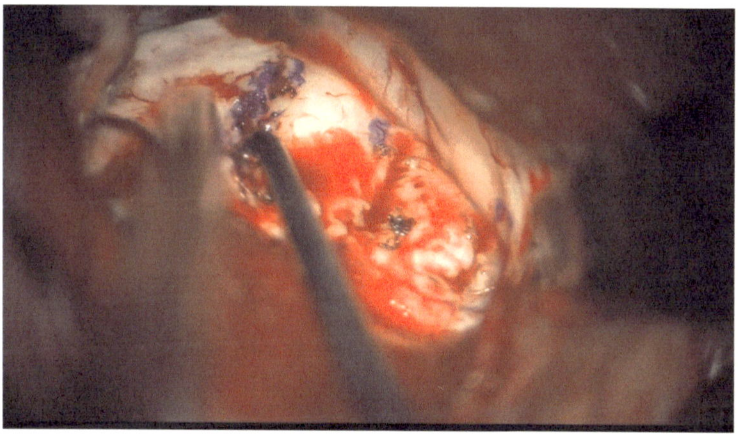

Figure 10.
After amygdalotomy multiple hippocampal transections (blue line) were performed to hippocampal head.

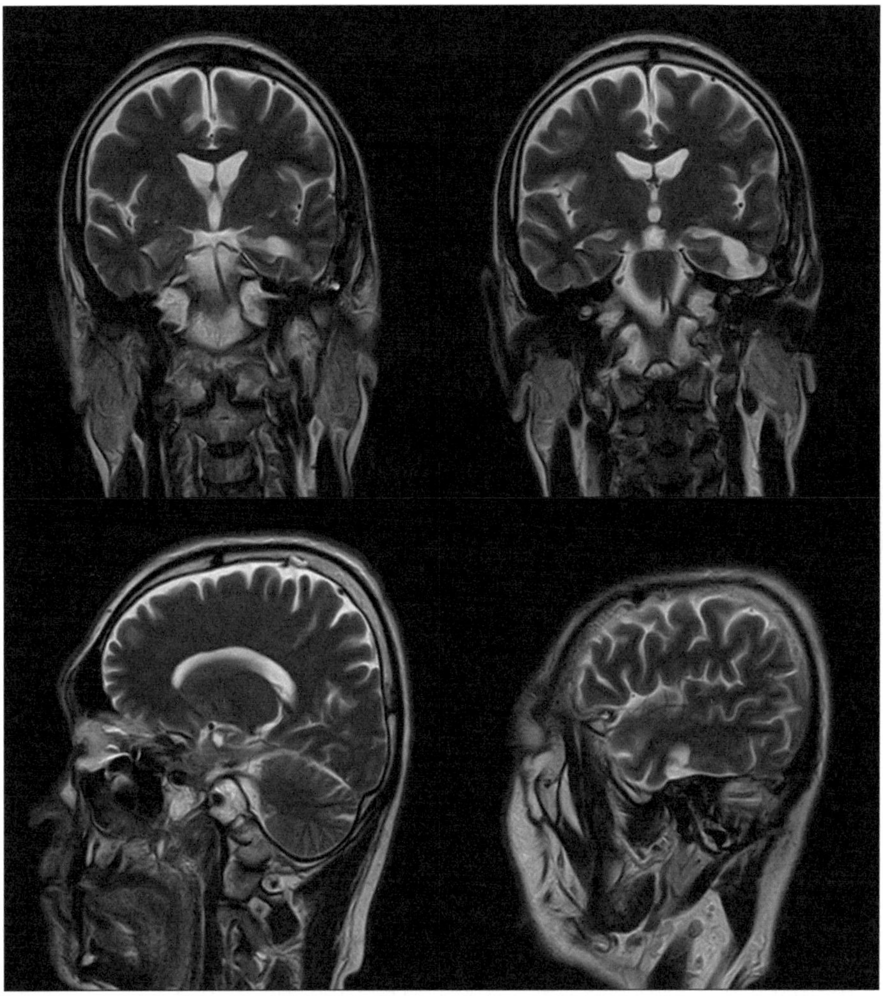

Figure 11.
Postoperative MRI demonstrated removal of basolateral nucleus of amygdala and preservation of hippocampus by MHT.

	2021/9/27: Ope date		
2021/7/1		2021/12/25	
VIQ	106	VIQ	119
PIQ	109	PIQ	137
TIQ	108	TIQ	129
VCI	105	VCI	111
PRI	121	PRI	146
WMI	113	WMI	113
PSI	102	PSI	118

VCI: verbal comprehension index, PRI: perceptual reasoning index, WMI: working memory index, PSI: processing speed index.

Table 4.
Pre- and postoperative neuropsychological examinations showing improvement in this patient.

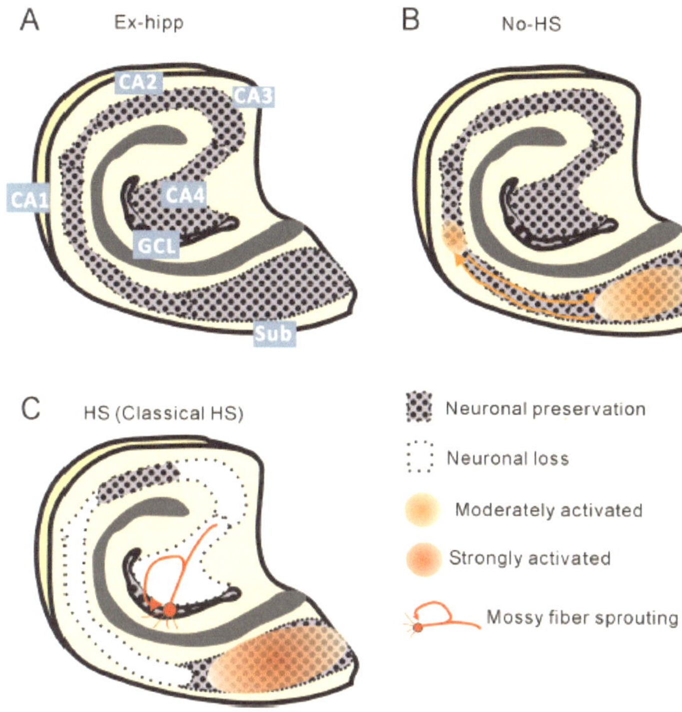

Figure 12.
Schematic representation of the results. (A) Control hippocampus showing anatomical orientation: from entorhinal cortex to granule cell layer (perforant fiber), from GCL to CA3 (mossy fiber), and from CA3 to CA1 (Schaffer collateral fiber). These fiber systems are called trisynaptic circuit of the hippocampus. Please refer to ***Figure 1****. (B) No-HS. Enhanced activities are initiated in the subiculum and extend backward to the CA1. Please refer to* ***Figure 8****. (C) HS. Epileptogenesis in the subiculum and MFS in GCL are evident. Adopted and modified from Kitaura et al. [20].*

results indicate that pathophysiological alterations involving the subiculum and dentate gyrus could be responsible for epileptogenesis in patients with MTLE (**Figure 12**).

In the No-HS group also, the activities in the subiculum and CA1 were temporarily correlated with each other. The activity in the subiculum is always being followed by that in CA1, suggesting that activity generated primarily in the subiculum was able to propagate into CA1 via feedback projection from the subiculum to CA1.

The activity generated primarily in the subiculum was able to propagate into CA1 via feedback projection from the subiculum to CA1. Kitaura et al. proposed that minimally invasive surgical approach involving disconnection of the circuit between the subiculum and the CA1 might be effective to control seizure.

11. New technique proposal for the language-dominant left TLE with HS (Type 1)

A 31-year-old woman is introduced to our clinic for the management of intractable left TLE. Her MR demonstrated typical HS. In consideration of her intractability, the amygdalotomy and multiple hippocampal transection by subtemporal approach with disconnection of subiculum and CA1 are considered to

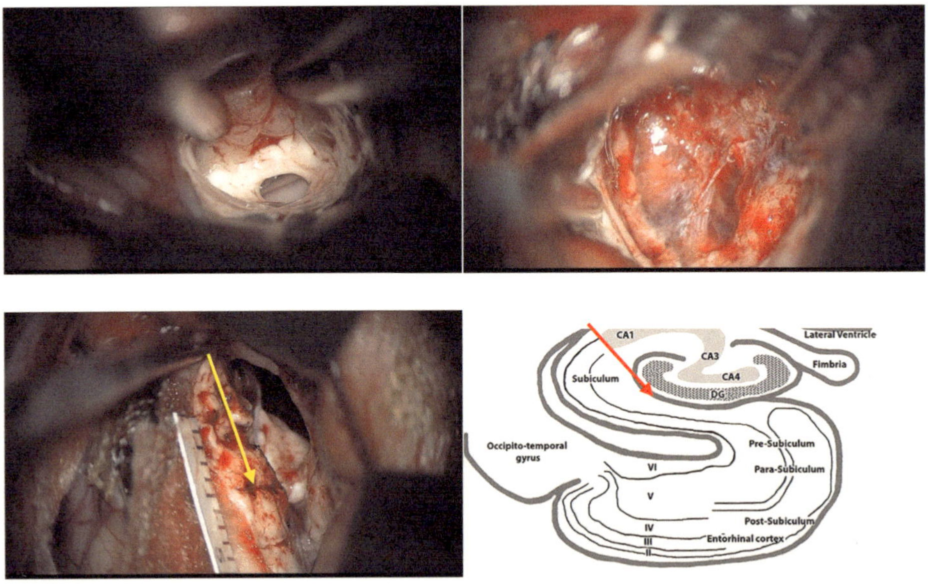

Figure 13.
Operative pictures showing amygdalotomy and hippocampal transection by subtemporal approach with disconnection of subiculum and CA1. Upper left: collateral sulcus (aspirator and forceps) and temporal horn opened showing hippocampus and amygdala. Upper right: After amygdalotomy (removal of basolateral nucleus), internal carotid artery and anterior choroidal artery were seen beyond arachnoid membrane. Lower left: After multiple hippocampal transection (blue line), continuous disconnection (yellow line) between CA1 and subiculum from alveus to hippocampal sulcus (lower right schema, red arrow) was done.

improve her seizure. Operative pictures show amygdalotomy and hippocampal transection by subtemporal approach with disconnection of subiculum and CA1 (**Figure 13**).

Postoperative neuropsychological examinations have improved already 1 day after the operation (Frontal Assessment Battery (FAB) and Hasegawa Dementia Scale (HDS)-Revised in **Table 5**), and seizure stopped. **Figure 14** demonstrated preoperative coronal T2W image and postoperative coronal T2W images,

Op. date: 2022/1/17.	
2018/4/27 (Preop.)	2022/3/30 (Postop. 2 months)
• WMS-R	• WMS-R
◦ Verbal Memory. 75	◦ Verbal Memory 78
◦ Visual Memory 87	◦ Visual Memory 89
◦ General Memory 74	◦ General Memory 78
◦ Attention/concentration 87	◦ Attention/concentration 81
• Delayed Recall 66	• Delayed Recall 83
◦ FAB 16/18	◦ FAB 18/18 (Preop. 14/18)
◦ TMT A 1′24″	◦ ADAS-J Cog 5.0/70
◦ TMT B 2′	◦ TMT A 35′
◦ MMSE 26/30	◦ TMT B 59”
	◦ HDS-R 30 (Preop 28)

Table 5.
Results of preoperative and postoperative neuropsychological examinations in this patient.

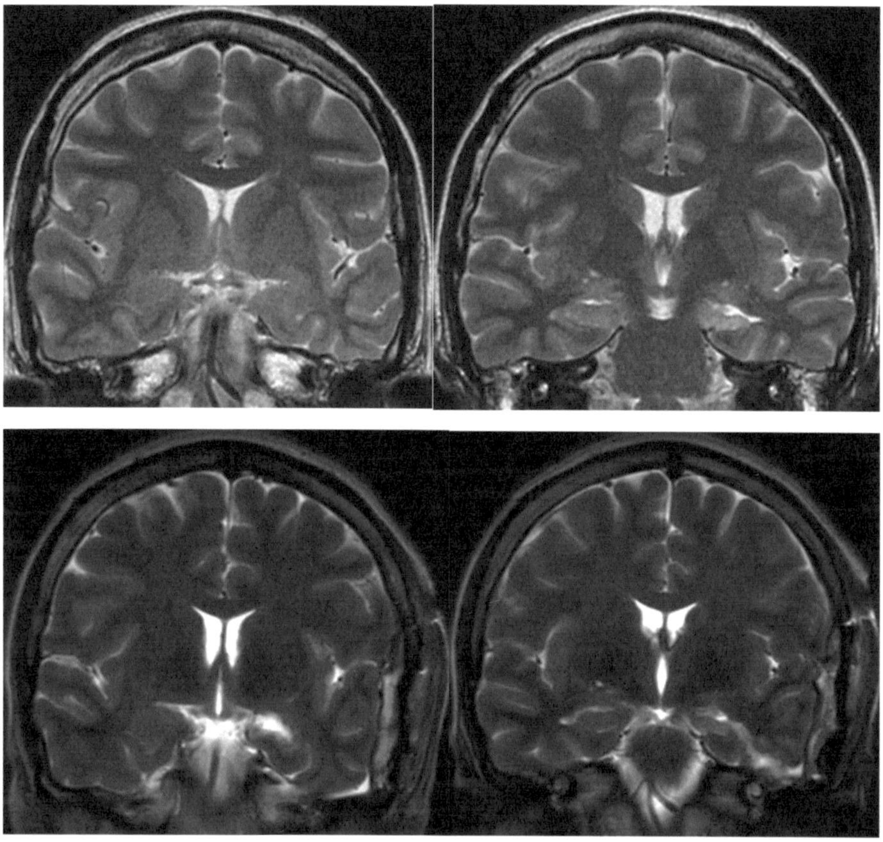

Figure 14.
Preoperative coronal T2W images showing amygdala (upper left) and HS (upper right), and postoperative coronal T2W images showing amygdalotomy (lower left) and MHT and disconnection between CA1 and subiculum.

showing amygdalotomy (lower left) and MHT and disconnection between CA1 and subiculum. Two months after surgery, neuropsychological examinations showed slight improvement comparing to preoperative levels (**Table 5** except FAB, HDS-R). Seizure control is satisfactory, and only two auras were seen during 1 year after the operation.

For disconnection between CA1 and subiculum for HS type 1, it is relatively easy to disconnect, because in HS type 1, CA1 is extremely atrophic and longitudinal limit of CA1 and subiculum is easy to identify.

12. Conclusion

1. Temporal lobe epilepsy, one of the most common forms of epilepsy involving limbic system and intractable for medical treatment, is the most challenging target for neurosurgeons, and various technical improvements are reported. Among these procedures, subtemporal selective amygdalohippocampctomy is seemingly least invasive in terms of neuropsychological function.

2. Technique and the operative and neuropsychological results of subtemporal selective amygdalohippocampectomy for medically intractable temporal lobe epilepsy have been demonstrated.

3. Using anatomical connectivity mapping technique, the subtemporal resection is the least disruptive to long-range connectivity which may explain its better cognitive outcome. These results have a direct impact on understanding the amount and/or type of cognitive deficit postsurgery, which may not be obtainable using local measures of white matter integrity.

4. New operative methods are demonstrated in this text, that is, subtemporal multiple hippocampal transection with or without disconnection between subiculum and CA1. These techniques might be useful to obtain good neuropsychological functions and good seizure control for language-dominant temporal lobe epilepsy.

5. Further operative experiences should be obtained to clarify the usefulness of subtemporal multiple hippocampal transections and disconnection of subiculum and CA1, especially for HS type 1, although randomized clinical trial is difficult to perform.

Abbreviations

TLE	temporal lobe epilepsy
TSSAH	transsylvian selective amygdalohippocampectomy
sSAH	subtemporal amygdalohippocampectomy
PTLE	paradoxical temporal lobe epilepsy
MHT	multiple hippocampal transection
TSMHT	transsylvianl multiple hippocampal transection
STMHT	subtemporal multiple transection

Author details

Tomokatsu Hori[1,2]*, Hideki Shiramizu[2] and Hajime Miyata[3]

1 Department of Neurosurgery, Tokyo Women's Medical University, Tokyo, Japan

2 Department of Neurosurgery, Moriyama Neurological Center Hospital, Tokyo, Japan

3 Department of Neuropathology, Research Institute for Brain and Blood Vessels-Akita, Akita, Japan

*Address all correspondence to: thori@moriyamaikai.or.jp

References

[1] Sekino Y, Obata K, Tanifuji M, Mizuno M, Murayama J. Delayed signal propagation via CA2 in rat hippocampal slices revealed by optical recording. Journal of Neurophysiology. 1997;**78**: 1662-1668

[2] Wiebe S, Blume WT, Girvin JP, Eliasziw M. Effectiveness and efficiency of surgery for temporal lobe epilepsy study group: A randomized, controlled trial of surgery for temporal-lobe epilepsy. The New England Journal of Medicine. 2001;**345**:311-318

[3] Cohen-Gadol AA, Bradley CC, Williamson A, Kim JH, Westerveld M, et al. Normal magnetic resonance imaging and medial temporal lobe epilepsy: the clinical syndrome of paradoxical temporal lobe epilepsy. Journal of Neurosurgery. 2005;**102**: 902-909

[4] Marathe K, Alim-Marvasti A, Dahele K, Xiao F, Buck S, O'Keeffe AG, et al. Resective, ablative and radiosurgical interventions for drug resistant mesial temporal lobe epilepsy: A systematic review and meta-analysis of outcomes. Frontiers in Neurology. 9 Dec 2021;**12**. Article 777845. www.frontiersin.org. DOI: 10.3389/fneur.2021.777845

[5] Morino M, Uda T, Naito K, Yoshimura M, Ishibashi K, Goto K, et al. Comparison of neuropsychological outcomes after selective amygdalohippocampectomy versus anterior temporal Lobectomy. Epilepsy & Behavior. 2006;**4**:95-100

[6] Shimizu H, Kawai K, Sunaga S, et al. Hippocampal transection for treatment of left temporal lobe epilepsy with preservation of verbal memory. Journal of Clinical Neuroscience. 2006;**13**:322-328

[7] Morino M. Surgical technique and neuropsychological outcome of transsylvian hippocampal transection in 26 patients with paradoxical temporal lobe epilepsy. Brain and Nerve (Japanese). 2011;**63**(4):347-354

[8] Takaya S, Mikuni N, Mitsueda T, Satow T, Taki J, Kinoshita M, et al. Improved cerebral function in mesial temporal lobe epilepsy after subtemporal amygdalohippocampectomy. Brain. 2009;**132**:185-194

[9] Hori T, Yamane F, Ochiai T, Kondo S, Shimizu S, Ishii K, et al. Selective subtemporal amygdalohippocampectomy for refractory temporal lobe epilepsy: operative and neuropsychologcal outcomes. Journal of Neurosurgery. 2007; **106**:134-141

[10] Usami K, Kubota M, Kawai K, Kunii N, Matsuo T, Ibayashi K, et al. Long-term outcome and neuroradiologic changes after multiple hippocampal transection combined with multiple subpial transection or lesionectomy for temporal lobe epilepsy. Epilepsia. 2016; **57**:931-940

[11] Hori T, Tabuchi S, Kurosaki M, Kondo S, Takenobu A, Watanabe T. Subtemporal amygdalohippocampectomy for treating medically intractable temporal lobe epilepsy. Neurosurgery. 1993;**33**:50-56

[12] Busby N, Halai AD, Parker GJM, Coope DJ, Ralph LMA. Mapping whole brain connectivity changes: The potential impact of different surgical resection approaches for temporal lobe epilepsy. Cortex. 2019;**113**:1-14

[13] Whiting AC, Chen T, Swanson KI, Walker CT, Godzik J, Catapano JS, et al. Seizure and neuropsychological

outcomes in a large series of selective amygdalohippocampectomies with a minimally invasive subtemporal approach. Journal of Neurosurgery. 2021;**134**:1685-1693

[14] Hori T, Miyata H, Seki T. Surgical results of temporal lobe epilepsy based on ILAE hippocampal sclerosis classification and abnormal immature granule cells are specific to hippocampal sclerosis Type 1. Annal Epilepsy Seizures. 2018;**1**:1-6

[15] Miyata H, Hori T, Vinters HV. Surgical pathology of epilepsy-associated non-neoplastic cerebral lesions: A brief introduction with special reference to hippocampal sclerosis and focal cortical dysplasia. Neuropathology. 2013;**33**:442-458

[16] Seki T, Hori T, Miyata H, Maehara M, Namba T. Analysis of proliferating neuronal progenitors and immature neurons in the human hippocampus surgically removed from control and epileptic patients. Scientific Reports. 2019;**9**:18194

[17] Uda T, Morino M, Ito H, et al. Transsylvian hippocampal transection for mesial temporal lobe epilepsy: Surgical indications, procedure, and postoperative seizure and memory outcomes. Journal of Neurosurgery. 2013;**119**:1098-1104

[18] Mikuni N, Miyamoto S, Ikeda S, Satow T, Taki Jn Takahashi J, et al. Subtemporal hippocampectomy preserving the basal temporal language area for intractable mesial temporal lobe epilepsy: Preliminary results. Epilepsia. 2006;**47**:1347-1353

[19] Miyamoto S, Kataoka H, Ikeda A, Takahashi J, Usui K, Takayama M, et al. A combined subtemporal and transventricular/transchoroidal fissure approach to medial temporal lesions. Neurosurgery. June 2004;**54**(5): 1162-1167; discussion 1167-1169. DOI: 10.1227/01.NEU.0000119234.61432.E8

[20] Kitaura H, Shirozu H, Masuda H, Fukuda M, Fujii Y, Kakita A. Pathoohsiological characteristics associated with epileptogenesis in hippocampal sclerosis. eBioMedicine. 2018;**29**:38-46

Chapter 9

Beyond Memory in H.M., The World's Most Famous "Hippocampal Amnesic"

Donald G. MacKay

Abstract

Patient H.M. (Henry Moliason) suffered a wide range of cognitive deficits attributable to the damage to his hippocampal formation but not to his well-established deficits in the ability to recall newly encountered facts, events, names, and objects, which formed the basis for his early diagnosis as a "hippocampal amnesic." Among Henry's "non-memory" or cognitive deficits that this chapter reviews are his impaired ability to create new and grammatical sentence plans, to identify who-did-what-to-whom in novel sentences and to understand and read aloud novel sentences containing familiar words that he understood without difficulty in isolation, but not in novel sentence contexts such as metaphors. Also reviewed are his deficits in detecting novel forms concealed within complex visual arrays, in identifying anomalous objects in novel visual scenes, in detecting and describing what makes captioned cartoons funny, and in combining familiar concepts into new and useful ideas. The chapter concludes by relating Henry's non-memory deficits to fundamental questions of this book, for example, What is the role of the hippocampal formation in human memory and cognition? And how does the hippocampal formation work?

Keywords: patient H.M., the hippocampal formation, language deficits, visual cognition deficits, memory deficits

1. Introduction

William Scoville discovered the link between memory and the hippocampal formation by chance in 1953 after he met Henry Moliason (age 27) in his office at the Hartford Hospital in Connecticut. Asked to eliminate or at least reduce the intensity and frequency of Henry's life-threatening epileptic seizures, Scoville surmised that Henry's convulsions probably originated in his hippocampus (in the middle of the brain, roughly between the ears: See **Figure 1**). With consent from Henry and his parents, Scoville drilled small holes in Henry's skull above his eyes, and using X-ray imagery, inserted thin metal tubes into the hippocampal region, and suctioned out about half of Henry's hippocampal formation bilaterally while leaving his neocortex virtually intact.

IntechOpen

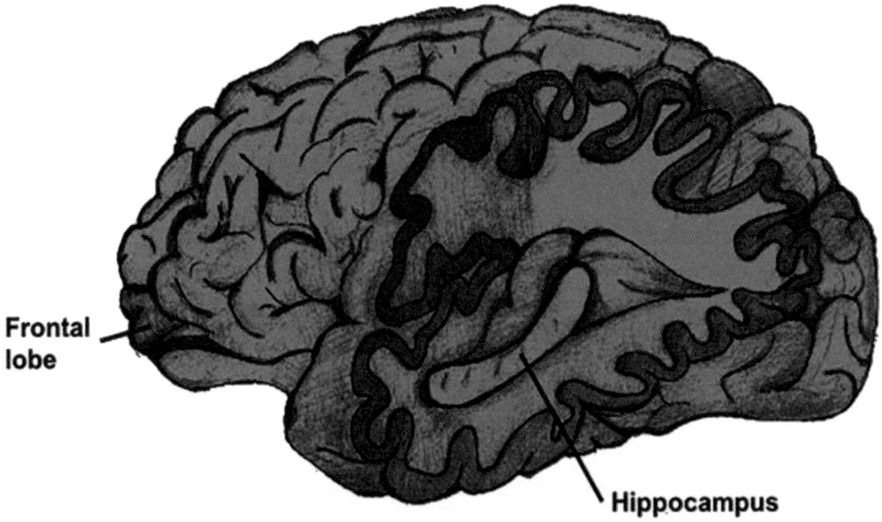

Figure 1.
The human hippocampus shown through a "window" into the medial temporal lobe. (Artist's rendition of an original illustration by Henry Vandyke Carter that appeared in Henry Gray's 1918 Anatomy of the Human Body.) Neither Henry's surgical damage nor the connections linking the hippocampus to surrounding structures in the hippocampal formation are shown.

This experimental operation essentially cured Henry's epilepsy and perhaps saved his life. However, it immediately became clear that something was terribly wrong. Henry could no longer remember things he had done hours, minutes, or even seconds earlier. He could not even find his way back from the bathroom to his bed in the hospital [1].

Henry's memory problems became a source of intense scientific scrutiny that soon made him famous around the world as "patient H.M." However, Scoville informed other neurosurgeons that he had inadvertently removed the engine for forming new memories in Henry's brain and warned against applying his surgical procedure in future cases.

Over the next fifty years, Henry participated in hundreds of psychological experiments at the Massachusetts Institute of Technology (MIT), and after he died at age eighty-two, neuroscientists conducted sophisticated analyses of his brain (which he bequeathed to MIT) and wrote several books about how research with Henry helped advance the behavioral and brain sciences [1, 2]. I dedicated my own recent book to Henry, "an ordinary man who became famous by generously devoting his life to helping scientists understand his memory, mind, and brain, trusting in the promise that what they learned about him would "help others." [3]. Researchers reading the present review of Henry's contributions to understanding the role of the hippocampal formation in human memory, visual cognition, language comprehension, and language production clearly represent the type of "others" that Henry wanted to help.

However, three limits to the scope of this review deserve comment. It only reviews studies that report deficits attributable to Henry's damaged hippocampal formation (rather than to damage elsewhere in his brain, e.g., the cerebellum), studies that experimentally study Henry's cognitive deficits independently of his memory deficits (by presenting, e.g., continuously displayed text, instructions, and pictorial stimuli that participants need not commit to memory), and studies that statistically compare performance for Henry versus control groups that resemble him in age, background, IQ, and education but lack damage in the hippocampal region. Also beyond the purview of this review are experiments with Henry not directly related to the memory versus non-memory issue (e.g., [4–11]), as well as the many experiments that established his famous deficits in recalling newly encountered names, facts, and events.

My UCLA lab conducted approximately twenty-five experiments with Henry at MIT and statistically compared his results in each study with those of 8–12 comparable control participants tested under similar conditions at UCLA.

2. Henry's non-memory deficits

This section reviews Henry's visual cognition, language comprehension, and sentence production deficits in studies that did not require the learning or recall of newly encountered names, facts, events, and objects (the classical definition of memory).

2.1 Henry's visual cognition deficits

Henry's performance on three tasks (hidden figure, what's-wrong-here, and visual cartoon comprehension) illustrate the general nature of his visual cognition deficits.

2.1.1 Henry's hidden figure deficit

Canadian neuropsychologist Brenda Milner accidentally discovered Henry's hidden figure deficit when trying to confirm her personal impression that Henry functioned without difficulty in his everyday visual world and only experienced problems when forced to remember where he had been or where he was going. To validate this impression, she had Henry complete the standard hidden figures test, a classic paper-and-pencil measure of the ability to detect target figures camouflaged in complex visual scenes. On a typical hidden figures trial, instructions asked Henry to inspect a shape (shown to the left on a page) and trace that target shape hidden within a complex concealing array shown to the right. In a real-life analog of this task, soldiers try to detect an enemy combatant (the target) camouflaged in a forest (the concealing array). Unlike a soldier, however, Henry had to detect abstract target shapes that were unmoving, unfamiliar to him, and visible throughout each trial (rendering recall unnecessary).

Henry correctly traced significantly fewer camouflaged targets than the memory-normal controls, a deficit indicating serious impairment to his ability to consciously recognize forms in complex visual scenes. However, Millner and many other neuropsychologists ignored Henry's hidden figures deficit and continued to consider him a "pure memory" case, with memory problems but no cognitive deficits [12].

Undeterred by this lack of interest, my UCLA lab replicated and refined Henry's hidden figures deficit by adding a condition where participants traced *familiar* hidden targets, which were forms that Henry experienced frequently in daily life before and after his lesion, e.g., squares, circles, right-angle triangles [13]. In this *familiar* target condition, Henry and the controls correctly traced the same number of targets, a seemingly minor fine-tuning of Henry's hidden figure deficit that paved the way for theoretical ideas discussed in Section 5 about why Henry experienced the selective deficits in cognition and memory that he did.

However, for the standard *unfamiliar* target condition, control participants in our study [14] traced reliably more targets in the concealing arrays than did Henry, a replication of Milner's hidden figure deficit that addresses a hypothesis in Bussey et al. that perceptual deficits due to parahippocampal damage reflect a memory problem [15]. Henry's hidden figure deficit in our study clearly reflected a perceptual problem [16].

2.1.2 Henry's what's-wrong-here deficits

Having confirmed and honed Henry's hidden figures deficit, my UCLA lab re-examined Henry's perceptual world in other ways. One involved a children's game found in books such as *What's Wrong Here: Hundreds of Zany Things to Find*. These books display complex everyday scenes, for example, a school classroom containing over a hundred busy people and objects, some of which are erroneous or anomalous, say, a bird flying upside down in a fish bowl filled with water, or an impossible-to-open door with hinges on the same side as its doorknob. Children enjoy discovering what's-wrong-here in pictures.

In our laboratory version of the game, Henry and suitable control participants inspected a series of what's-wrong-here pictures, circled as many erroneous objects as possible within each picture, and explained why each circled object was anomalous within a generous time limit [17].

Henry correctly circled significantly fewer erroneous objects than the controls, and he misidentified many of his circled objects without correcting himself, something the controls never did. For example, Henry called a clearly drawn rabbit "a dog" and called an ordinary wastebasket on the floor beside a teacher's desk "a window." Because Henry always correctly identified the identical objects *when depicted in isolation in a subsequent test*, Henry's object identification errors were clearly specific to the what's-wrong-here scenes, consistent with a problem in disentangling unfamiliar forms from their unfamiliar surroundings in complex visual displays, as in the original hidden figure test [18].

2.1.3 Henry's cartoon comprehension deficits

New higher-level deficits in Henry's visual cognition and sentence comprehension emerged in an experimental test of his ability to understand captioned cartoons [19]. Participants in this UCLA study saw a sequence of cartoons with instructions to explain what made them funny and to read aloud their captions (which only contained words Henry knew before his surgery). One example is Gary Larson's cartoon, "Raising the dead," which depicts two women in armchairs chatting —a normal scene except that both are *ghosts*. A ghost woman named Edith is listening to the other complain about problems raising her ghost children, Billy and Sally. Illustrating her difficulties, the cartoon shows Sally floating head first down a stairway, while Billy flits aimlessly around the room. The caption reads: "Oh, I don't know. Billy's been having trouble in school and Sally's always having some sort of crisis. I tell you, Edith, it's not easy raising the dead."

Control participants in this study consistently detected humor in the cartoons and never misread their captions in ways that would preclude comprehension of why they were funny. Not Henry. For example, Henry misread the Larson caption as, "I tell Edith it's not [long pause] easy, the raising the dead," messing up the sentence prosody and omitting the critical word *you* in *I tell you, Edith, it's not easy raising the dead,* all without self-correction. Henry apparently thought the speaker was talking to someone not depicted in the cartoon, rather than to Edith, the ghost mother seated beside her.

Henry's uncorrected caption-reading errors did not just render full understanding of cartoons impossible. They also suggested an inability to grasp *who-said-what-to-whom* in the cartoons, reflecting a serious comprehension deficit that my UCLA lab established in a subsequent study (described shortly).

Nor was confusion about who spoke to whom Henry's only problem. He did not grasp what the cartoons were about, another basic prerequisite to getting the jokes. For example, Henry did not see that the Larsen cartoon depicted ghosts. Noting that he could see through Larson's ghost-speaker to the armchair on which she sat, Henry suggested that the cartoonist had drawn her wrong and complained that "she" (the cartoonist) just "bl. .. the. .. blackens the whole way, and everything.. ." using some kind of "blackening rule." Unlike Henry, the control participants never misidentified visual forms in the cartoons nor mistakenly ascribed an outline to artistic error.

3. Henry's language comprehension deficits

Consistent with Henry's caption-reading errors, four sources of evidence indicated deficits in Henry's ability to comprehend and read aloud various types of sentences, a skill he mastered in grade school and high school, many years before his age 27 surgery.

3.1 Deficits in identifying who-did-what-to-whom in sentences

Henry's failure to comprehend who-did-what-to-whom in Larson's cartoon echoed an important finding in a 1966 experiment that I conducted at MIT [20]. In that study, Henry saw various types of ambiguous sentences on cards with instructions to describe the two meanings of each sentence as quickly as possible. Henry readily detected both meanings in some types of ambiguous sentences, but relative to controls, he displayed a major deficit in detecting the dual meanings of sentences resembling *John is the one to help today,* where *John should help us* is one meaning and *we should help John* is the other. Henry clearly had a problem working out who-did-what-to-whom in ambiguous sentences.

But could Henry understand who-did-what-to-whom in *unambiguous* sentences? To find out, my UCLA lab ran an experiment in which Henry and memory-normal controls read unambiguous sentences on a computer screen, one at a time, and then answered a multiple-choice comprehension question displayed on the same screen [21]. For example, after reading *The water that the mother spilled surprised the young child*, participants answered the comprehension question *who spilled the water: the mother, the young child, or nobody*? Control participants correctly answered significantly more comprehension questions than Henry, firmly establishing a deficit in his ability to understand the most important information that sentences can convey: who-does-what-to-whom [22].

3.2 Deficits in reading sentences aloud

Can Henry comprehend and accurately read aloud the individual words in unambiguous sentences? To find out, my UCLA lab first had participants read lists of familiar words presented one at a time on cards, for example, *GOT, ATE, STOMACH, HOT, WHO, DOGS, ACHES, BOYS, and THE*. Henry made no more mistakes than the controls when reading those isolated words.

Days later, however, Henry experienced major deficits when asked to read the same words re-organized into sentences, for example, *the boys who ate hot dogs got stomach aches,* instead of GOT, *ATE, STOMACH, HOT, WHO, DOGS, ACHES, BOYS, and THE.*

Unlike the controls, Henry now made dozens of uncorrected reading errors that rendered his utterances ungrammatical. He also paused abnormally at critical points within the sentences, for example, misreading *The boys who ate hot dogs got stomach aches* as *The boys* [unusually long pause] *ate hot dogs got stomach aches* (Note Henry's ungrammatical omission of the word *who*) [23].

3.3 Deficits in comprehending metaphors

Metaphors, such as *Life is a Journey,* are powerful linguistic tools. They shape everyday thinking and help people comprehend and learn ideas that are otherwise difficult to acquire [24]. Can Henry comprehend metaphors? To answer this question, my UCLA lab asked Henry and suitable controls to indicate what metaphoric sentences mean. On each trial, participants saw a short metaphoric sentence on a computer screen with instructions to choose the best of three possible ways to interpret it. By way of illustration, these were the three choices for the metaphor *Maybe we should stew over his suggestion*:

1. *Let us think about it some more* (correct metaphoric interpretation),

2. *Maybe we should put more meat into his idea* (incorrect but metaphoric interpretation),

3. *Let us make sure to cook the stew long enough* (incorrect literal interpretation containing the same critical word *stew* as the target sentence).

The memory-normal controls chose the correct metaphoric interpretation reliably more often than Henry, indicating a deficit in his ability to comprehend metaphors. Indeed, Henry performed worse than chance (random guessing) because he usually chose the incorrect interpretation with the same critical word as the target sentence; here the word *STEW* capitalized in the original metaphor *Maybe we should STEW over his suggestion* and the incorrect literal interpretation: *Let us make sure to cook the STEW long enough* [23]. Henry clearly understood the *individual words* but not the overall meaning of the metaphoric sentences.

3.4 Deficits in detecting what's right versus wrong in sentences

Another UCLA study tested whether Henry could distinguish between grammatical versus ungrammatical sentences. He could not. Asked to respond "Yes," to grammatical sentences such as *She hurt herself,* and "No," to ungrammatical sentences such as *He hurt herself*, Henry answered correctly significantly less often than suitable control participants, with performance close to chance (50%). Henry clearly had a deficit in comprehending whether sentences are grammatical versus ungrammatical [25].

Did he fail to understand the instructions? Did he not care? To find out, my UCLA team reran the previous study, adding foil sentences and a test for guessing. The foil sentences thoroughly shuffled the words in grammatical sentences such as *She has decided to buy a house*, yielding ungrammatical strings such as *Decided has house she a buy to*. Like the control participants, Henry invariably called these foils ungrammatical, indicating clear comprehension of the instructions.

To assess guessing, the experimenter immediately asked participants who responded, "No, ungrammatical," to identify the wrong or misplaced word and then correct that word to make the sentence grammatical. This was easy for the control participants. For example, after identifying *be* as the misplaced word in *Will be Harry blamed for the accident*, they quickly produced a corrected version such as *Will Harry be blamed for the accident.*

Not Henry, however. He called correct words incorrect and failed to correct words he deemed wrong. For example, Henry identified *blamed* as the incorrect word in *Will be Harry blamed for the accident*, but insisted that further information about the blame was needed to correct this error. Henry was indeed guessing when he called sentences ungrammatical.

4. Henry's language production deficits

Many sources of evidence indicate that Henry suffered language production deficits. Length constraints limit us to just one source here: His performance on the standardized test of language competence (TLC) [26]. In a typical TLC trial, participants see two words above a picture, together with continuously displayed instructions to use both words in a single grammatical sentence that accurately describes the picture.

Control participants found this task easy. For example, asked to use the words *ALTHOUGH* and *WRONG* in a single grammatical sentence that describes a woman in a sports store discussing a tracksuit with a salesman, one control participant quickly responded *The woman decided to buy the suit ALTHOUGH it looked WRONG.* A panel of judges blind to speaker identity later rated this transcribed response 100% correct on the three evaluative dimensions shown in **Table 1**.

Table 1 also shows Henry's response to the same sportswear picture: "*Because it's wrong for her to be he's dressed just as this that he's dressed and the same way*," a response that the panel of judges rated as inaccurate, ungrammatical and incoherent, and a rambling series of non sequiturs (see **Table 1**). Across all trials, Henry included significantly fewer must use words than the controls, and the panel of judges rated his utterances ungrammatical, inaccurate, or incoherent significantly more often than those of the controls.

Why did Henry include significantly fewer must use words in his utterances than the controls? The coherence rating for Henry's utterance in **Table 1** suggests one reason. Henry was freely generating familiar phrases (e.g., *the same way*) without relating the picture to the must use words, a free association strategy that may also explain why the judges more often considered Henry's TLC descriptions incoherent, ungrammatical, and inaccurate.

Did the damage to Henry's hippocampal formation shape his free association strategy? Almost certainly. Henry could easily retrieve phrase memories formed before his age 27 surgery, for example, the common phrase *the same way*. However,

Scene description	**A woman in a sports store is discussing a tracksuit with a salesman**
Must use words (in caps throughout)	ALTHOUGH WRONG
100% correct Response of a control participant	*"The woman decided to buy the suit ALTHOUGH it looked WRONG."*
Henry's response to the same scene	*"Because it's WRONG for her to be he's dressed just as this that he's dressed and the same way."*
Henry's must use words score	50% correct due to omission of the must use word ALTHOUGH
Henry's accuracy evaluation	INACCURATE: No two people in the scene are dressed "the same way."
Henry's grammaticality evaluation	UNGRAMMATICAL: Henry's utterance is not a sentence.
Henry's coherence evaluation	INCOHERENT: a rambling series of non sequiturs.

Table 1.
An illustrative TLC trial, with scene description, must use words, the completely correct response of a control participant, and Henry's response, rated on three evaluative dimensions and scored for inclusion of the must use words.

the damage to his hippocampal formation prevented him from forming new and situation-appropriate phrases and sentences. How do we know? Because my UCLA lab analyzed hundreds of unintended and uncorrected errors that rendered Henry's TLC utterances ungrammatical, inaccurate, and incoherent errors that speak volumes about how the hippocampal formation goes about creating new memories in the cortex.

Henry's TLC errors fell into two categories: *Omissions* (where participants omit units that are essential in a grammatical sentence) and *Combination errors* (where participants conjoin two or more units into a sequence that is impermissible or ungrammatical). **Table 2** illustrates both types of error in an utterance Henry produced on a single TLC trial. The TLC picture shows three people: a woman server at a cafeteria counter, a man ordering food from her, and a woman ahead of him in line who already has the food she ordered on a tray. *PIE* and *EITHER* are the must use words.

To describe this scene, a typical control participant produced both must use words in a single grammatical sentence, for example, "*I want* either *some* pie *or some cake*" (see **Table 2**). Not Henry, however. Instead of his intended utterance, *I want some of what she had*, Henry said "I want some her [long pause] what she had" (see **Table 2**).

How did this study determine what participants intended, planned or wanted to say? Determining what *normal participants* intended to say after they made an error was easy. Our experimenters simply asked them what they meant or noted how they spontaneously corrected their errors. For example, a normal speaker who says *Put it on the chair, I mean table*, clearly intended to say, *Put it on the table*. These scoring procedures indicated that control participants occasionally produced omission errors on the TLC but made no category-combination errors whatsoever.

For the hundreds of errors that Henry produced on the TLC, however, determining intent was more challenging because Henry never spontaneously corrected his omission and category-combination errors and never clarified what he meant to say when asked [27]. My UCLA lab therefore developed and adopted a more general set of

Scene description	**A man is ordering food at a cafeteria counter, and a woman ahead of him in line already has her food on a tray.**
Must use words (in caps throughout)	PIE and EITHER
Henry's actual response	"I want some her. .. what she had."
Henry's intended response	*I want some of what she had.*
Omission error analysis (omitted word *of* in parentheses in Henry's intended utterance)	*I want some (of) what she had.*
Combination error analysis (ungrammatical word combination *some her* in caps and italics)	"I want *SOME HER.*"
Actual response of a typical control participant	"I want EITHER some PIE or some cake."
intended response of that control participant	*I want EITHER some PIE or some cake.*

Table 2.
Omission and combination errors illustrated in a single TLC trial, with scene description, must use words, and Henry's errors analyzed by comparing his actual versus intended utterance and the correct response of a typical control participant.

scoring procedures that allowed us to specify participants' intent (independent of the speaker) as the "best possible correction" of an anomalous utterance (see [28]).

Henry's missing word *of* is clearly an omission error that renders his utterance ungrammatical (see **Table 2**). However, why is his phrase "*SOME HER*" ungrammatical? The reason is that only common nouns (e.g., *fun* and *games*) can follow an indefinite determiner such as *some* in grammatical English sentences (e.g., *We played some games and had some fun*). When the pronoun *her* follows *some* in a phrase, an utterance becomes ungrammatical.

Another important observation about Henry's *SOME HER* is that the word *HER* intrudes some aspect of the upcoming word *SHE* in Henry's intended utterance, *I want some of what SHE had*. What aspect of *SHE* intruded? It was not its syntax because unlike *SHE*, *HER* is a possessive pronoun. It was not its speech sounds because *SHE* and *HER* share no speech sounds whatsoever.

Rather Henry's *HER* almost certainly reflects intrusion of the *CONCEPT* "female," which underlies three aspects of what he was trying to say: the forthcoming word *SHE* in his intended utterance, the lady server in the TLC picture, and the woman leaving with food on her tray, an analysis suggesting that Henry's TLC errors may reflect a breakdown in the uniquely human ability to combine conceptual units when creating situation-appropriate sentences such as *I want some of what she had*.

The next section expands on this idea, arguing that Henry's errors lay bare the sophisticated and elegant functions of the neural machinery that allow the human hippocampal formation to conjoin smaller concepts into larger internal representations in the cortex including internal representations for comprehending, perceiving, remembering and describing experiences, and events and the visual world.

5. The hippocampal formation in cognition and memory: Lessons from H.M

What possible lessons can the Brain and Cognitive Sciences take from the research with Henry reviewed here? This section outlines five categories of lessons: 1).

Different brain mechanisms create new memories versus retrieve old or preformed ones, 2). Distinguishing between pre-formed versus newly formed memories in the brain can be tricky, 3). The hippocampal formation performs the same basic function in visual cognition, language comprehension, and language production, 4). And performs the same basic function in memory for facts, names, events, and common objects, and 5). Lessons from future tests of hypotheses derived from research with Henry reviewed here.

5.1 Distinct mechanisms create new memories vs. retrieve old ones

Why did Henry misread the sentence, The boys who ate hot dogs got stomach aches, as the boys ate hot dogs [abnormally long pause] got stomach aches (omitting the critical word WHO), whereas he easily and correctly read the same words presented one at a time in isolation, for example, GOT, ATE, STOMACH, HOT, WHO, DOGS, ACHES, BOYS and THE?

This type of finding (repeated across every domain of cognition that we have examined with H.M.) indicates that mechanisms in the hippocampal formation create new internal representations, whereas a separate mechanism located elsewhere retrieves old or pre-formed neural representations. For example, current evidence indicates that memories for familiar words reside in the language areas of the neocortex whereas mechanisms for retrieving those words reside in the frontal lobes. Because Henry's frontal lobes and cortical language areas were intact, he could, therefore, retrieve and read without difficulty isolated words learned before his surgery.

However, word retrieval is insufficient to make novel sentences sound like sentences when reading aloud. Engaging the hippocampal formation to create new internal representations of the relations between words and phrases in novel sentences is necessary to do this. For example, to correctly read the sentence *The boys who ate hot dogs got stomach aches,* the hippocampal formation must create three new neocortical phrase units to represent *the boys, ate hot dogs*, and *got stomach aches*, and to signal the relations between them by adding the word *who* and inserting pauses of varying lengths, as in *The boys* [short pause] *who ate hot dogs* [major pause] *got stomach aches*. The damage to Henry's hippocampal formation, therefore, prevented him from doing this.

Nevertheless, Henry deserves thanks for *trying* to read the sentences as sentences. He might have adopted a word-by-word reading strategy throughout our reading studies, pausing after each word in the sentences as if reading a list. This strategy would have precluded the mysterious pauses and word omissions that my lab was at pains to explain. Because Henry did not adopt this strategy, science now has a clear understanding of how the hippocampal formation contributes to normal sentence reading.

5.2 New versus old memories in the brain: Lessons from H.M.

The distinction between new versus old or preformed memories in the brain was a source of confusion in early research with Henry. For example, Dr. Brenda Milner, the famous Canadian neuropsychologist, defined any never previously encountered stimulus as *new*, an assumption that led her to falsely conclude that the hippocampus plays no role in processing new perceptual information. To refine Milner's new versus old definition and demonstrate the critical role of the hippocampal formation in novel perceptual processing required decades of research.

To see why, consider in detail the Gollin fragmented-figures test of perceptual abilities that Milner administered to Henry and memory-normal controls in the 1960s. On the first trial of this test, participants see a picture of a familiar object, say, an elephant, that is so fragmented that nobody can correctly guess what it is. In subsequent trials (2–5), participants see the same picture with progressively less fragmentation until everyone can correctly identify version 5. The participant's goal is to correctly name all of the fragmented objects in as few trials as possible.

The results indicated that Henry could initially identify a fragmented image as readily as controls, even though that exact fragment pattern had not been viewed before. Her conclusion: Henry's hippocampal lesion did not prevent him from processing new perceptual information.

But were the fragmented pictures *as overall stimulus patterns* really the basis for participants' responses? It seems more likely that they correctly guessed, for example, "elephant," as soon as a fragment in the progressively less fragmented picture of an elephant revealed a unique elephant feature, say, its distinctive tusk, trunk, or tail. If so, Henry's non-deficit merely indicates what the present research has shown: that retrieving and recognizing visual features that Henry acquired long before his lesion does not require hippocampal engagement. Based on his childhood experiences with elephants and elephant pictures, Henry could respond "elephant" with no need to create a new internal representation for the complete fragmented elephant picture *per se*.

We can, therefore, return to the original question: Does hippocampal engagement play a role in processing new perceptual information? Results from two other conditions in Milner's fragmented-figure study suggest that maybe it did. One involved a simple rerun of the test one hour later. When Milner's participants again saw the same progressively less fragmented pictures repeated, performance improved significantly more for the memory-normal controls than for Henry. Why? Milner suggested that the normal controls achieved this benefit by learning the verbal labels of the Gollin figures during the first test, allowing them (but not Henry because of his verbal memory deficits) to quickly sample from the correct population of names on the retest. However, another retest 20 weeks later contradicted this name recall hypothesis. Although interference should have obliterated Henry's memory for the names by then, Henry performed better on the hidden figure test after the 20-week delay than after the one-hour delay.

Finally, Milner's results do not contradict a plausible alternate hypothesis that the intact hippocampal system of the normal participants created new internal representations of the evolving perceptual information on the fragmented-figures test so that they (but not Henry) could remember how, say, fragments of the elephant's trunk evolved from unrecognizable to recognizable as the elephant picture became progressively less fragmented, thereby enabling faster correct recognition of the objects *per se* (and not just their names).

In summary, the distinction between new versus old in cognition and the brain is subtle, multidimensional, and dependent on the *functional* stimuli in a task. The functional stimuli can be new when normal participants initially experience a sequence of hidden figures but not when they experience the same sequence a second time. Similarly for reading isolated words versus sentences. To read isolated words, list-like prosody (fixed pause durations between the words) suffices, but instructions to correctly read a novel sentence creates a functionally different stimulus that requires speakers to compute the relations between words in the sentence and adjust their prosodic intonation and pause lengths accordingly.

Another important dimension to the new versus old distinction is the state of a participant's memories. To count as old rather than new, a stimulus must have an internal representation in the participant's brain that is *pre-formed and functional for the task at hand.* In a lexical decision task, for example, where participants must respond YES to words and NO to nonwords, a once familiar but now forgotten word can represent a new rather than old stimulus if aging and infrequent use has degraded the participant's internal representation for that word (see e.g., [29]).

5.3 The hippocampal formation functions similarly across different cognitive domains

Despite obvious differences in how language comprehension, sentence planning, and visual cognition are tested, Henry's deficits indicate that the hippocampus serves to create new internal representations in all three domains. For example, Henry's deficit in the standard hidden figures test indicated that lacking a hippocampus, he could not form the new internal representations required to detect unfamiliar targets in concealing arrays. However, he readily detected familiar targets, for example, squares, circles, and right-angle triangles, because he acquired pre-formed internal representations of those target forms long before the lesion to his hippocampal formation.

Henry's deficits in detecting anomalous objects in what's-wrong-here scenes, for example, an impossible-to-open door with hinges on the same side as its doorknob, demands a similar account. For Henry, an impossible-to-open door looks normal because, lacking an intact hippocampal formation, he could not form a new internal representation of the novel relations between hinges and doorknob that distinguish normal from impossible doors.

Henry's language comprehension deficits require a similar account. For example, grammatical sentences such as *She hurt herself* and ungrammatical sentences such as *He hurt herself* were equivalent for Henry because, without a functional hippocampal system, he could not form new internal representations of the relations between the words in either type of sentence. Similarly for metaphors, Henry's damaged hippocampal system prevented him from creating the new internal representations required to comprehend one kind of event, for example, *taking the time to talk and think about something*—in terms of another— *cooking slowly, as with a stew* in the metaphoric sentence *Maybe we should stew over his suggestion*.

Similarly in language production. Why did Henry produce hundreds of ungrammatical utterances on the TLC, saying, for example, "I want some her [long pause] what she had," when asked to use two continuously displayed words in a single grammatical sentence describing a picture of a man, a cafeteria counter, and a woman with food on a tray? The answer is that without an intact hippocampal formation, Henry could not relate the TLC picture to the must use words in order to create a new internal representation for producing grammatical sentences such as *I want either some cake or some pie.*

5.4 The hippocampal formation functions similarly in cognition and the classical domains of memory

To compare how the hippocampal formation functions in cognition (previous section) versus the four classical domains of memory, this section examines the role of the intact hippocampal system in creating memories for newly encountered facts, names, events, and objects.

Memory for facts. How would a young child form a memory for 2x2 = 4 as a newly encountered fact? Via hippocampal engagement that creates a new internal representation resembling a sentence that means *Two multiplied by two is four*.

Memory for names. How would normal speakers of English create an internal representation of the newly encountered name of my son: *Ken MacKay*? Via hippocampal engagement that simultaneously and powerfully activates two preformed units in the cortex, one representing his familiar given name, *KEN*, and the other representing his family name, MACKAY (familiar to anyone reading this chapter). The powerful co-activation of these preformed units will quickly create strong new synapses that link *KEN* and *MACKAY* to a new or "uncommitted" neural unit that will represent his combined first and last names.

However, weak new connections can also be formed *without hippocampal engagement* when preformed units are repeatedly activated over prolonged periods of time. This explains why, for example, Henry slowly came to recognize and occasionally use the name *Suzanne Corkin* after encountering her name virtually daily over decades, one of many observations suggesting that normal hippocampal engagement simply speeds up the fundamental process of massive repetition that underlies all new connection formation.

Memory for events. How would a normal adult form memories of recently experienced events such as a *visit to the dental clinic*? Via hippocampal engagement that creates an internal representation that conjoins neural units in event categories such as [actor] + [action] + [where] + [when], much like the hippocampal engagement process that creates sentences such as *I stupidly scheduled my dentist for that day* or *He happily clobbered the ball out of the stadium:* by conjoining neural units in the linguistic categories [pronoun] + [adverb] + [verb] + [noun phrase] + [prepositional phrase].

Memory for common objects. How do children create memories for frequently encountered objects such as a classic American penny? By engaging their hippocampus to form an internal representation that is good enough to distinguish pennies from other coins and objects. This good enough internal representation consists of a surprisingly small number of perceptual features, for example, *small, round, copper-colored,* and *engraved with the profile of Abraham Lincoln* that children then use to guide their subsequent interactions with pennies [30, 31].

As a consequence, naturally acquired penny memories are quite unlike an eidetic image in the brain that one might inspect and report as an adult. Such an eidetic image would include at least 37 penny features resembling those in **Table 3** below (all of which are easy to see in the photographs of a penny shown in **Figure 2**).

So children's ability to recall only three or four features of a penny represents an accuracy level of about 10%, and *adult* participants in memory experiments, e.g., [32], achieve a *similar accuracy level*, reflecting virtually no improvement relative to children. Why do decades of everyday interactions with pennies yield so little learning? The reason is that adults only rarely, if at all engage their hippocampal formation to add new penny features to their "good enough" internal representation of a penny that they formed as children and have continued to use in everyday financial transactions since then.

5.5 Possible lessons from future tests of hypotheses derived from research with Henry

New lessons for the field may come from future tests of hypotheses derived from the research with Henry reviewed here. To illustrate just one of many such testable hypotheses, consider the claim in Section 5.1 (on Memory for names) that engagement

1. *GENERAL, small*
2. *round,*
3. *copper-colored.*
4. *FRONT SIDE, ABRAHAM LINCOLN* PROFILE,
5. rightward-facing,
6. curved at bottom.
7. DATE,
8. below chin,
9. chest level,
10. small font.
11. LIBERTY,
12. caps,
13. small font,
14. behind profile,
15. neck level.
16. *IN GOD WE TRUST,*
17. caps,
18. medium font,
19. centered above head,
20. curved.
21. Cravat tucked into shirt.
22. BACK SIDE: Lincoln Memorial,
23. centered on coin.
24. E PLURIBUS UNUM,
25. with dots e·pluribus·unum·,
26. above the Lincoln Memorial,
27. all caps,
28. small font.
29. ONE CENT,
30. caps,
31. below Memorial,
32. curved,
33. large font.
34. UNITED STATES OF AMERICA,
35. curved,
36. caps,
37. medium font.

Table 3.
Thirty-seven Features of a Classic American Penny, with major features in caps and subordinate and minor features in lower case. To verify the features, see the photographs in ***Figure*** *2.*

Figure 2.
Photographs that verify the 36 penny features analyzed in ***Table 3.***

of activating mechanisms in the hippocampal formation serves to simultaneously and powerfully activate two preformed units in the cortex, thereby quickly creating strong new synapses that link both preformed units to a new or "uncommitted" neural unit that constitutes the internal representation for a newly encountered name such as *Ken MacKay*. A future study employing advanced technology will be able to test whether two preformed units in the cortex become simultaneously and powerfully activated when participants learn a newly encountered combination of familiar proper names. That same study will also be able to determine whether the strong co-activation of those preformed cortical units originated somewhere within the hippocampal formation. And the study that reports both of these hypothetical results will feature H.M. in its reference section. So will a possible follow-on study that precisely localizes where in the hippocampal formation the mechanisms for co-activating proper names are located.

6. Conclusions

In addition to his well-known deficits in memory for newly encountered names, events, facts, and objects, H.M. experienced a wide range of non-memory deficits reviewed here. Four conclusions emerged: 1). The hippocampal formation creates new internal representations in the cortex for comprehending novel linguistic information, perceiving novel visual forms, and creating novel sentences, 2). The hippocampal formation likewise creates new internal representations for freshly encountered facts, names, events, and objects, the classical domains of memory, 3). The hippocampal formation does not store preformed memories, nor is it essential for their retrieval. Mechanisms for retrieving preformed internal representations from the cortex reside elsewhere in the brain, for example, the frontal cortex, 4). Finally, Henry's contributions to the Brain and Cognitive Sciences seem unlikely to end soon as future studies continue to test hypotheses derived from research with Henry, especially recent hypotheses about how the hippocampal formation works in memory, visual cognition, language comprehension, and sentence production.

Author details

Donald G. MacKay
Psychology Department, University of California Los Angeles, CA, USA

*Address all correspondence to: mackay@ucla.edu

References

[1] Corkin S. Permanent Present Tense: The Unforgettable Life of the Amnesic Patient, H.M. New York: Basic Books; 2013. p. 363

[2] MacKay DG. Remembering: What Fifty Years of Research with Famous Amnesia Patient H. M. Can Teach us about Memory and How it Works. New York: Prometheus Books; 2019. p. 400

[3] MacKay DG. Remembering: What Fifty Years of Research with Famous Amnesia Patient H. M. Can Teach us about Memory and How It Works. New York: Prometheus Books; 2019. p. 6

[4] James LE, MacKay DG. Word knowledge and aging: Support for a new theory of long-term retrograde amnesia. Psychological Science. 2001;**12**:485-492. Also available at http://mackay.bol.ucla.edu/publications.html

[5] MacKay DG, Stewart R, Burke DM. H.M. revisited: Relations between language comprehension, memory, and the hippocampal system. Journal of Cognitive Neuroscience. 1998;**10**:377-394

[6] MacKay DG, Burke DM, Stewart R. H.M.'s language production deficits: Implications for relations between memory, semantic binding, and the hippocampal system. Journal of Memory and Language. 1998;**38**:28-69. Also available at http://mackay.bol.ucla.edu/publications.html

[7] MacKay DG, James LE. Aging, retrograde amnesia, and the binding problem for phonology and orthography: A longitudinal study of "hippocampal amnesic". Aging, Neuropsychology and Cognition. 2002;**9**:298-333. Also available at http://mackay.bol.ucla.edu/publications.html

[8] MacKay DG. Aging, memory and language in amnesic H.M. Hippocampus. 2006;**16**:491-495. Also available at http://mackay.bol.ucla.edu/publications.html

[9] MacKay DG, Hadley C. Supra-normal age-linked retrograde amnesia: Lessons from an older amnesic (H.M.). Hippocampus;**19**:424-445. Also available at http://mackay.bol.ucla.edu/publications.html

[10] MacKay DG, Goldstein R. Creativity, comprehension, conversation and the hippocampal region: New data and theory. AIMS Neuroscience. **3**:105-142. Available at http://www.aimspress.com/journal/neuroscience

[11] MacKay DG. The earthquake that reshaped the intellectual landscape of memory, mind and brain: Case H.M. In: Macpherson SE, Della Sala SD, editors. Single Case Studies of Memory. New York: Routledge; 2019. pp. 16-39

[12] Milner B. Memory and the temporal regions of the brain. In: Pribram KH, Broadbent DE, editors. Biology of Memory. New York: Academic Press; 1970. pp. 29-50

[13] MacKay DG, James LE. Visual cognition in amnesic H.M.: Selective deficits on the what's-wrong-here and hidden figure tasks. Journal of Experimental and Clinical Neuropsychology. 2009;**31**:769-789. Also available at http://mackay.bol.ucla.edu/publications.html

[14] MacKay DG, James LE. Visual cognition in amnesic H.M.: Selective deficits on the what's-wrong-here and hidden figure tasks. Journal of Experimental and Clinical Neuropsychology. 2009;**31**:769-789. Also

available at http://mackay.bol.ucla.edu/publications.html

[15] Bussey TJ, Saksida LM, Murray EA. Impairments in visual discrimination after perirhinal cortex lesions: Testing "declarative" vs. "perceptual mnemonic" views of perirhinal cortex function. European Journal of Neuroscience. 2003;**17**:649-660

[16] MacKay DG, James LE. Visual cognition in amnesic H.M.: Selective deficits on the what's-wrong-here and hidden figure tasks. Journal of Experimental and Clinical Neuropsychology. 2009;**31**:769-789. Also available at http://mackay.bol.ucla.edu/publications.html

[17] MacKay DG, James LE. Visual cognition in amnesic H.M.: Selective deficits on the what's-wrong-here and hidden figure tasks. Journal of Experimental and Clinical Neuropsychology. 2009;**31**:769-789. Also available at http://mackay.bol.ucla.edu/publications.html

[18] MacKay DG, James LE. Visual cognition in amnesic H.M.: Selective deficits on the what's-wrong-here and hidden figure tasks. Journal of Experimental and Clinical Neuropsychology. 2009;**31**:769-789. Also available at http://mackay.bol.ucla.edu/publications.html

[19] MacKay DG. Remembering: What Fifty Years of Research with Famous Amnesia Patient H. M. Can Teach us about Memory and How it Works. New York: Prometheus Books; 2019. pp. 233-336

[20] MacKay DG. The Search for Ambiguity by an Amnesic Patient: Implications for the Theory of Comprehension, Memory, and Attention. Psychology Department, University of California Los Angeles; 1972 [Unpublished MS]

[21] MacKay DG, James LE, Taylor JK, et al. Amnesic H.M. exhibits parallel deficits and sparing in language and memory: Systems versus binding theory accounts. Language and Cognitive Processes. 2007;**22**(3):377-452. available at http://mackay.bol.ucla.edu/publications.html

[22] MacKay DG, James LE, Taylor JK, et al. Amnesic H.M. exhibits parallel deficits and sparing in language and memory: Systems versus binding theory account. Language and Cognitive Processes. 2007;**22**(3):377-452. available at http://mackay.bol.ucla.edu/publications.html

[23] MacKay DG, James LE. The binding problem for syntax, semantics, and prosody: H.M.'s selective sentence-reading deficits under the theoretical-syndrome approach. Language and Cognitive Processes. 2001;**16**:419-460

[24] Materials were from Wiig E.H.M. and Secord W. The 'Thematic Roles' Subtest, Test of Language Competence. New York: The Psychological Corporation, Harcourt, Brace, Jovanovich; 1988

[25] MacKay DG, James LE. The binding problem for syntax, semantics, and prosody: H.M.'s selective sentence-reading deficits under the theoretical-syndrome approach. Language and Cognitive Processes. 2001;**16**:419-460. Also available at http://mackay.bol.ucla.edu/publications.html

[26] MacKay DG, James LE, Hadley CB, Amnesic HM. Language production, and memory: Selective deficits on the Test of Language Competence. Journal of Clinical and Experimental Neuropsychology. 2008;**30**:280-300. Also

available at http://mackay.bol.ucla.edu/publications.html

[27] MacKay DG, James LE, Hadley CB, Amnesic HM. Language production, and memory: Selective deficits on the Test of Language Competence. Journal of Clinical and Experimental Neuropsychology. 2008;**30**:280-300. Also available at http://mackay.bol.ucla.edu/publications.html

[28] MacKay DG, Johnson LW. Errors, error detection, error correction and hippocampal-region damage: Data and theories. Neuropsychologia. 2013;**2013**:2633-2650. DOI: 10.1016/j.neuropsychologia.2013.08.009

[29] James LE, MacKay DG. Word knowledge and aging: Support for a new theory of long-term retrograde amnesia. Psychological Science. 2001;**12**:485-492. Also available at http://mackay.bol.ucla.edu/publications.html

[30] McCurdy HG. Coin perception studies and the concept of schemata, Word knowledge and aging: Support for a new theory of long-term retrograde amnesia. Psychological Review. 1956;**63**(3):160-168. DOI: 10.1037/h0046614

[31] Kikuno H. Schema of coins and memory for features of coins. Perceptual and Motor Skills. 1993;**77**(3):1235-1238

[32] Nickerson RA, Adams MJ. Long term memory for a common object. Cognitive Psychology. 1979;**11**:287-307